Malicorne

Malicorne

Earthly Reflections of an Astrophysicist

HUBERT REEVES

Translated from the French
by Donald Winkler

Stoddart

First published in English in 1993 by
Stoddart Publishing Co. Limited
34 Lesmill Road
Toronto, Canada
M3B 2T6
(416) 445-3333

French language edition first published in 1990 by
Editions du Seuil

Canadian Cataloguing in Publication Data

Reeves, Hubert, 1932-
Malicorne

Translation of: Malicorne.
ISBN 0-7737-2689-6

1. Science — Philosophy. I. Title.

Q175.R3513 1993 501 C92-095635-1

Cover design by Brant Cowie/ArtPlus Limited
Typeset by Tony Gordon Ltd.
Printed and bound in Canada

Stoddart Publishing gratefully acknowledges the support of the Canada Council, Ontario Ministry of Culture and Communications, Ontario Arts Council and Ontario Publishing Centre in the development of writing and publishing in Canada.

To lovers of science
and poetry

Contents

Malicorne

A June Morning in Malicorne

I LIFT MY HAND to blot out the morning sun, and a delicate dance of pollen is revealed against the blue. Tiny white tufts emerge from shadow, drift slowly on currents of air, glimmer for an instant, then vanish into the leaves to form new patterns in the darkness. Each tuft carries a seed, precious cargo, a powerful databank it will deposit somewhere in the countryside that new springtimes may flourish.

At my feet, a stream flows. Large alders line the banks of its bed, and the water, its eddies silvered by the sun, threads and churns its way past knotted roots. Shimmering patterns waver and gleam in thin, luminous bands. A dragonfly, blue with a bronze sheen, hovers noisily.

What a pleasure to see these alders, with their networks of trunks issuing from a tangle of half-submerged roots, with their heavy branches angled over the stream bed, an array of bridges, and with silvered circles floating from out of their shade on the water's surface. The sad and tireless song of a turtledove reaches my ear.

Sky mirrored in water, shifting reflections of blue among boughs languidly swaying: I love this landscape. My steps lead me naturally toward the shaded streams that flow deep in the valleys. I sit on a tree trunk. To preserve the fleeting memory of these treasured moments, I record them on a tape machine. The lines you've just read have fixed in my mind a glorious morning in June on the banks of the Branlin.

Malicorne is a small village in Puisaye, in the north of Burgundy, home to the French writer Colette. In this lush countryside, green and gently rolling, one can walk for hours without meeting a single automobile. It's a place where you must take off your watch and adapt to the rhythms of nature, from which we've been so cruelly cut off by modern life. Here one is aware of the changing seasons. The odor of acacias or lime trees marks the springtime, and the appearance of purple crocuses in the wet fields tells us, nonchalantly, that autumn has already arrived.

This book was born of my rambles in this countryside. It almost wrote itself, behind my back. I was more spectator than author. I owe it to the trails of golden light on a shining carpet of periwinkles in the shade of the trees.

Here, I am at peace. Alert to sounds and smells, I wake to the tranquil presence of the natural world. I feel alive, on the surface of planet Earth, at this precise moment in the evolution of the universe.

As I stroll, images come to mind, and thoughts that have been ripening for some time deep inside. In the quietness of this place they leap into my consciousness, sometimes as a simple, fairly succinct utterance. At other times, words in torrents spring forth and flow along, unstoppable. I dictate them just as they come, into my faithful tape recorder. Back at the house, I transcribe it all into a notebook. Over the years, I've accumulated hundreds of pages.

What to do with all this material? I first thought of publishing my notebook just as it was written, like a diary. But I soon abandoned that idea. In its original form, it would appear inexcusably confused and incoherent. On the other hand, to extract from it a solidly structured text would frustrate my wish to retain what is subjective and personal in these pages.

I've adopted a compromise solution. I've itemized my nagging preoccupations, the themes that turn up most frequently. Then I've grouped together the texts that, to one degree or another, correspond to these themes. So as not to lose that sense of spontaneity I consider so important, I've done nothing to avoid ambiguities and repetitions. Above all, I did not want to "philosophize," in the traditional sense of the term. I distrust the coherence of synoptic thinking and the logic of "world systems." Each of us tries to work through his or her own life and destiny. Perhaps these reflections will be of service to some, as both testimony and inspiration. That, at least, was my hope in writing this book.

Now that the work is done, I can describe it more easily. Implicit throughout is the question of how science relates to culture, the relationship between what one *knows* and what one *does*. I have asked myself how new discoveries tend to influence the way we look at various spheres of human activity. The book's chapters deal successively with the connections between science and poetry, science and human freedom, science and artistic creation, science and the workings of justice, science and religion.

Of course, this concern is not new. One's vision of the world evolves along with scientific progress. The observation of planets through a telescope, for instance, contributed to the cultural revolution we call the Renaissance. The Earth was no longer at the center of the cosmos. Once people became aware of the immensity of the universe, all modes of thought and human activity were affected. The impact was felt even in daily life. Bertolt Brecht's *The Life of Galileo* dramatizes this eloquently.

It's the observation of galaxies, at the beginning of the twentieth century, that led to the most important discovery of contemporary science: our universe has a history! In *Atoms of Silence*, I outlined the facts and arguments that support this new world picture.

To quote Jean-Marc Lévy-Leblond, we are always "introducing science into culture." How does the concept of a historical universe affect the way we think and act? This is not a matter of seeking simplistic "explanations" for human activity. Jurisprudence, artistic creation, and mystic rapture exist, each in its own right, and have no need for scientific justification. More modestly, we may hope that, by seeing them in the context of the history to which they belong, we may eventually shed new light on these activities.

The theme of "reconciliation" will be present throughout these pages. In the first part, "The Play of Light," we will see how the scientific and poetic visions, far from being mutually exclusive, work together to help us see the world in all its splendor.

The evolution of scientific thought as it relates to "freedom" and "creativity" is the subject of Part Two: "Butterflies over a Field of Colza." Thanks to developments in mathematics and astronomy, these notions are no longer foreign to scientific theory.

The third section groups together reflections on a variety of human activities. Artistic creation, legislation, and religious practice take on new connotations when seen in the light of current knowledge. Here, for our purposes, we'll have to alter the popular expression "from the point of view of Sirius" to "from the point of view of fossil radiation." On the scale that concerns us, Sirius is just not far enough away . . .

Frequently, in these pages, we will talk of *play*. We will analyze the role of play in the complex unfolding of the cosmos. At the same time, we'll try to define the preconditions for the *zones of freedom* where this play can occur, both in the heart of the cosmos and within a human being.

More than once, we'll find ourselves in a charmed kingdom, *the playground*, where the small child comes face to face with the outside world. The British psychoanalyst D. W. Winnicott has pointed out the vital importance of this environment to various activities of the human psyche. For each one, but in different ways, it's a source of energy and motivation. To study these activities in this context is to glimpse their underlying unity and their specific differences. Here, age-old conflicts and apparent contradictions begin to resolve themselves.

The title of this book is a reminder that here there are no structured and coherent theories, but rather personal impressions, inspired by the simplest of phenomena: the play of the sun's light on a tree, the forest floor in flower under the first warm rays of the spring sun, a flight of butterflies, a steeple, a smokestack in the countryside. By setting these reflections in rural surroundings, I wanted to provide myself with a "zone of freedom" — a playground! — from where I might venture onto perilous terrain, as one can among friends.

Above all, I didn't want to lose the personal flavor of these reflections. Like any other human being, I'm contained in the reality I describe. As I grapple with the conflicts and problems of human life, I react with my own temperament and emotions. They have left their traces everywhere in these pages.

A word on the vocabulary. I make no real distinction between the words "nature," "reality," and "universe." Moreover, I have a regrettable tendency to personalize these phenomena. I would have been happy if the publisher had distributed, along with each book, a bag of quotation marks. Then each reader could insert them, according to his tastes, wherever he saw fit. Although it's best not to take the anthropomorphic language too seriously, as a teaching tool it's extremely helpful . . .

At Malicorne, we live among the great trees of an old Burgundian farm. I adore planting trees, particularly those with a long life expectancy: cedars of Lebanon, Himalayan cedars, metasequoias of Szechwan, red oaks, chestnut trees. I've planted sequoias that are now over five meters high. In our Malicornian jargon, this group of trees is called the "millenary forest."

It gives me great satisfaction to feel I'm responsible for these trees, which will long survive me. (Should someone take it into his head to chop them down, I shall rise from my grave.)

I pay them visits. Their strong presence speaks to me and inspires me. I acknowledge with pleasure their contribution to the reflections in this book.

PART ONE

The Play of Light

Science and Poetry

1

The Empire of Numbers

THE RAW HUES AND PASTELS of autumn foliage stand out against a vast blue sky. Among the dark conifers and scarlet maples, the aspens' glossy leaves quiver in the morning breeze. Rapid movements, rhythmic oscillations, like the gleaming, silvery reflections of light on the sea.

Framed by the white trunks, this trembling, or rather this swarming of brilliant light, evokes in the viewer a feeling of fullness and perfection. Everything here is in its proper place.

But appreciation is not enough for the human spirit. It also wants to understand. Today, after two thousand years of scientific research, we have learned much. We know why the sky is blue or green or pink and why the leaves change color. We know the laws that govern the graceful fluttering of leaves at the ends of their stalks. We can calculate the number of photons the foliage reflects into our dazzled eyes.

But is not the magic of this scene threatened by the completeness of our explanations? Has poetry still something to say once science has done its work? Can the elusive charm of this fall morning stand up to the analysis of these delicate mechanisms, laid bare by the inexorable eye of the researcher?

This issue crops up on a regular basis during the question

periods that follow my lectures. It reflects a widespread contemporary concern. If I broach the subject in these pages, it's because it is at the crossroads of my own natural interest in science and poetry, and I have often considered the problem myself.

SUNSET OVER THE PACIFIC

IT ALL BEGAN WHEN I was eighteen years old. I was then on the west coast of Canada for a summer apprenticeship at the Dominion Astrophysical Observatory.

In the evening, I often went to see the sun set over the ocean. The incident I'm about to relate took place during a particularly magnificent twilight, whose every detail, after forty years, is still engraved in my memory.

Far off, the snowcapped summits of the coastal mountains were turning slowly from white to pink. Reflected in the calm ocean waters, the colors spread themselves over the long swells moving in from the open sea.

As I was lost in meditation, a sudden thought broke my mood and "wrenched me from my reverie like a tooth," as Jacques Prévert would have said.

I'd been familiar with maritime scenes for a long time. Over and over again, I had roamed the rugged Gaspé coasts and the broad beaches of the Gulf of St. Lawrence. But since my last visit to the ocean something important had happened. As a student in the physics department at the University of Montreal, I had encountered, some months earlier, Maxwell's equations.

Maxwell was a Scottish physicist of the nineteenth century.

To him we owe, in large part, the theory of light. His equations provide us with an excellent mathematical representation of light's behavior. Reflection, refraction, diffraction, interference: all this play of light originates in the interaction of atoms with electric and magnetic fields. Superb, mathematically elegant, and functional, Maxwell's theory elicits an enthusiastic response from any student coming across it for the first time. It is one of the great moments of apprenticeship in physics.

As I contemplated this calm ocean, gloriously tinted by the setting sun, an inner voice spoke: "These designs, these forms, these shimmering hues are the mathematical solutions to Maxwell's equations. Perfectly predictable and calculable. Nothing more."

Within, I panicked. I feared that the exquisite pleasure I had enjoyed would simply dissipate. Must I renounce it forever now that I'd peeked over the fence and sampled the poisoned fruit of knowledge? As unthinkable as this cruel sacrifice seemed, it was equally impossible to turn back the clock. In my mind, on the horizon, Maxwell's equations stood tall, cold, and inexorable. Their harsh light cancelled out, it seemed, the fragile magic of the rose-tinted sky and the iridescent sea.

Shaken by this quandary, I turned my back on a panorama I could no longer bear and walked home alongside the giant sequoias of Beacon Hill Park. A violent headache assailed me, like a sharp piece of metal embedded in my skull. Trapped in this dilemma, I understood for the first time the root meaning of the word "schizophrenia": to be torn apart, mentally.

That evening left its mark on me for a long time. It was the beginning of a lengthy journey that continues to this day. In search of a solution, or rather a reconciliation, I came to explore a number of avenues. I set off down roads I never intended to take in hope of reclaiming the right to enjoy in peace the sight of pink waves on a tranquil sea.

It is this route that I would like to retrace for those who, like me, at a certain point in their lives have felt the pain caused by this sharp piece of metal. Can we still surrender ourselves to the spectacle of the natural world once scientific thought has allowed us to peer backstage?

At the outset, let me give fair warning. This inquiry is best regarded as an impromptu tour, or the sort of account one gives to friends on returning from a trip abroad. Together we'll explore many spheres of human activity. My goal is not to arrive at a definitive synthesis. We'll retain only those elements that will help us to light our way.

It is common practice, before embarking on the recital of a long journey, to give one's listeners an overview of the territory to be covered. This is especially valuable when the itinerary involves twists and turns, detours, and steps retraced. It makes it possible, at any moment, to get a feeling for the overall picture.

Four chapters will be devoted to this odyssey. The first will retrace the birth and ascendancy of the empire of numbers, while the second will analyze the causes of its decline. We shall see how, after establishing itself bit by bit as the absolute and complete expression of reality, mathematics encountered certain difficulties that cut it back down to size. Logic, the foundation of mathematics, appears to us now as an evolving process, an imperfect undertaking, like all human activity, desperately seeking its own coherence. In its new station in life, it has had to forego those pretensions to supremacy that it is no longer in a position to sustain.

In the next two chapters, we'll try to understand how logical thought made its appearance in human history. We'll seek to identify its basic elements (Chapter Three). Psychoanalysis will help us locate its point of origin: a small child's anxiety, faced with a frightening reality (Chapter Four).

Seen in a more modest and realistic context, the true role of logical thought is not, as was sometimes claimed, to supplant all other overtures of the human spirit, but to work with them in order to grasp and explore together the world's many dimensions. "The full mind is alone the clear," writes Schiller. In other words, all human faculties must be brought into play if we're to deal with reality in all its subtlety and complexity.

THE EMPIRE OF NUMBERS

WHERE DOES THIS claim to superiority that I so willingly granted to scientific discourse come from? By what right can it sweep away all other approaches? How can it demand — a much more serious offence — that I deprive myself of a boundless pleasure: the naive appreciation of natural beauty?

The story of the empire of numbers begins more than 2,500 years ago on the shores of the Mediterranean. Trying to comprehend the world around them, Greek thinkers had a brilliant idea. Behind the extraordinary complexity of visible reality, there must exist an invisible "simplicity." This "simplicity" was composed of a few fundamental elements — bricks, of a sort — out of which everything was built in an ordered and organized fashion.

To "conceive" reality required one first to identify these elements, then to demonstrate, one by one, the mechanisms according to which our complex, visible reality formed itself from these primordial building blocks.

This quest for reality's basic bricks continued for several centuries. In the beginning, various material elements were proposed: earth, air, fire, water.[1] These concrete substances were soon replaced by more abstract elements. There was a progression from the world of things to that of ideas.

PYTHAGOREAN MYTHOLOGY

WE ALL REMEMBER the Pythagorean Theorem concerning the sum of the squares of the sides of a right-angle triangle. But we know very little about Pythagoras himself. It's said that, convinced from childhood of the fundamental importance of numbers, he went to Egypt to study mathematics with the priests of the Pharaonic cult, just as today one might seek higher knowledge in the great American universities.

Pythagoras was enlightened by music. There's a simple relationship between the length of a lyre's strings and the sound produced. Halving the string raises the pitch by an octave. Reducing it by two-thirds produces a fifth; by three-quarters, a fourth. In the same way, the sound produced by a hammer on the anvil is proportional to the weight of the hammer.

Inspired by the musical harmony of hammers and vibrating strings, Pythagoras set forth a proposition that was revolutionary for his time: *nature is fundamentally mathematical.* Numbers govern all of reality. The key to the cosmos is the numeral.

Pythagoras and his disciples set out to apply this beautiful idea to all spheres of human activity, including morality and justice. It was at one and the same time a synthesis and a formidable research program.

"What is the wisest?" asked the Pythagoreans. "The number." "And what is the most beautiful?" "The harmony of numbers." It's from this period that we derive the expression "music of the spheres." According to some accounts, one member of the school, Eurytus, set himself the task of determining the *essential number* of every living being. To do so he would calculate, for example, the number of small stones needed to compose the image of a man or a horse.[2]

This way of thinking was strongly reinforced by Plato. For him, *ideas* play approximately the same role as *numbers* for Pythagoras. Material reality is an illusion. Nature, ultimately, is made up of ideas. These ideas "exist" in an abstract "otherworld," from where they create and control all aspects of our universe. Pythagoras and Plato agreed, "God is a geometer."

THE LANGUAGE OF GOD

THIS POINT OF VIEW enjoyed enormous success, dominating the development of Western thought and science. Over the centuries, it was echoed by an impressive number of distinguished thinkers.

Galileo wrote: "Philosophy, written in that great book . . . the universe . . . is written in the mathematical language . . . without whose help it is impossible to comprehend a single word of it; without which one wanders in vain through a dark labyrinth." "Mathematics is the language of God," said Newton. And Einstein: "The world is ultimately intelligible in terms of geometry." Richard Feynman, one of the greatest physicists of our time, readily confessed in private conversations to his belief in the ultimate reality of ideas and numbers.[3]

Bertrand Russell, an early twentieth-century British mathematician, is for us a key player in this story. He experienced firsthand the triumph and decline of the empire of numbers. At the start of his career, he venerated mathematical theorems. For him, they were at once Plato's "ideas," Spinoza's "eternal order," and the "world of being" of all the metaphysicians. Carried away by his enthusiasm, he sings the science of numbers in prose that is frankly lyrical: "The world of being is unchangeable, rigid, exact, delightful to the mathematician, the logician, the builder of metaphysical systems, and all who love perfection more than life." "Mathematics, rightly viewed, possesses not only truth but supreme beauty — a beauty cold and austere, like that of sculpture, without appeal to any part of our weaker nature, without the gorgeous trappings of painting or music, yet sublimely pure, and capable of a stern perfection such as only the greatest art can show."[4]

These lyrical passages from a great mathematician — he will later have a change of heart — entitle us to speak of a "Pythagorean mythology" to describe this absolute faith in ideas and numbers that prevailed for over two thousand years.

"Visionaries of the transcendent," scoffed Nietzsche, lucid and sarcastic, referring to the proponents of this faith. We will see that events were largely to prove him right. For the moment, however, let's keep our eyes on the spectacular expansion of the mathematical empire.

THE LADDER OF KNOWLEDGE

OVER THE CENTURIES, numbers gradually made inroads into all fields of knowledge.

Galileo (it is said) rolled balls down inclined planes, then measured the distance traveled in terms of the time taken. He applied these results to the study of moving bodies. Thus he swept away, in one stroke, two thousand years of Aristotelian physics. Absolute movement did not exist; absolute space did not exist. Both were relative to the observer.

Pascal recorded the changes in barometric readings between the town of Clermont-Ferrand and the tops of the surrounding mountains. From this, he deduced the existence of the earth's atmosphere.

Descartes measured the angles at which light reflects off mirrors. From these observations he derived the fundamental laws of geometric optics.

For years, Tycho Brahe noted with precision the positions of the planets in the sky. Thanks to these data, Johannes Kepler showed that planetary orbits are not circles but *ellipses*.

Planetary ellipses are *almost* circles; the discrepancies are minor. Kepler, said his contemporaries, is wasting his time. But here — as is so often the case in scientific research — a numerical difference, even a small one, "makes all the difference."

The implications were enormous. Thanks to the work of Tycho Brahe and Kepler, Newton was able to develop his theory of universal gravitation. Since then, the movements of the planets and the tides hold no mystery. We can calculate, millennia in advance, the positions of the sun and moon.

The empire of numbers continued its triumphant progress. It went beyond the limits of astronomy and invaded all of physics. The mysteries of light were clarified by Maxwell's electromagnetic theory. The properties of heat were revealed by thermodynamics. Lavoisier and his successors extended the reign of numbers to the field of chemistry. *And so, at this point, it was concluded that numbers were the common language of science*.

Naturally, each scientific discipline had its own concepts and methods, its own way of looking at reality. But even if chemistry, physics, and mathematics were self-contained sciences, they were far from independent. Chemistry, for instance, borrowed its concepts from physics, and physics borrowed its from mathematics. Even biology was studied in terms of organic chemistry.

A PEDESTAL OF PURE BRONZE

SUCH WAS THE SITUATION at the end of the nineteenth century, when the French philosopher Auguste Comte proposed a structure for the empire of numbers. He established a "ladder of the sciences." At the top of the ladder he put "human sciences" such as sociology and psychology, which relate to biology. But biology is based on chemistry, which itself is based on physics, and physics is built on mathematics.

Thus all spheres of knowledge rested one upon the other and were ultimately grounded in that pedestal of pure bronze that was numbers and logic.

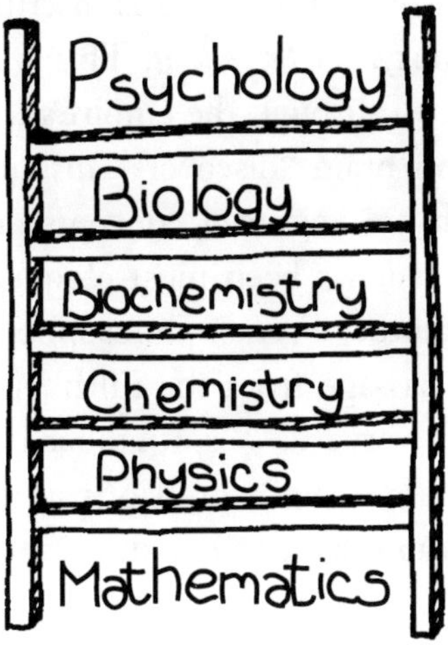

Auguste Comte's ladder, 1875.

Comte's ladder of sciences, synthesizing two thousand years of scientific research, confirmed Pythagoras's brilliant hunch. Eternal and all-powerful, mathematics, for a long line of researchers, represented true reality.

HOW DID MATHEMATICS COME TO HUMANKIND?

At THIS POINT, one question comes naturally to mind. How does this world of numbers make itself known to the human mind? How do we learn about it? What mental mechanisms allow the brain to grasp it? It's a question that will haunt us for the rest of this chapter. It will permeate our investigation of science and poetry.

For Plato, the world of numbers is inscribed, unbeknownst to us, in our memory. *We "recall" it.* This is the initial answer, and it will prevail all through the empire's long rise.

That the human brain "discovers" inside itself the "stone tablets" on which are engraved concepts, logic, and mathematics is a view that has been most eloquently expressed by the seventeenth-century French philosopher René Descartes.

I quote here a passage from the Fifth Metaphysical Meditation, which will serve as a springboard for a number of reflections. I've put certain expressions in italics, in order to emphasize their importance to our discussion:

> On first discovering [mathematics] it seems that I am not so much learning something new as remembering what I knew before; or it seems like *noticing for the first time things which were long present within me* although I had never turned my mental gaze upon them before.
>
> But I think the most important consideration at this point is that I find within me countless ideas of things which *even though they may not exist anywhere outside me* still cannot be called nothing . . . but have their own true and immutable natures.[5]

In the age of Christian metaphysics, the reality of numbers seemed perfectly natural. For Descartes, as for the scholastic philosophers, "truth" existed in God. Mathematics sprang from the same source. The divine commandments were transmitted to Moses on tablets of stone. The laws of numbers were engraved from birth in human memory. We had only to learn to read them.

This belief survived the secularization of philosophic thought in subsequent centuries. We find it still in a number of contemporary thinkers.

To end this chapter, let us sum up our investigation, and the ground we have covered so far. Does science, in providing an explanation for sunsets, kill their magic? This is the question that started us off. We took ourselves back to the time of the Greek thinkers. There we encountered a thesis that, for thousands of years, has enjoyed great success. Numbers are the ultimate reality. They exist prior to everything we perceive. They are inscribed in our brain. Distinguished scientists are convinced that in discovering the mathematical laws of the cosmos, they are reading in the great book of God's design.

2

The Decline of the Empire

DURING THE NINETEENTH CENTURY, the view of mathematics as something engraved in human memory and underpinning all that is real would suffer a series of setbacks. In this second chapter, we'll describe them one by one. In particular, we'll see how, under their impact, the ladder of sciences was transformed into a "circle of knowledge." This circle, so vicious as to be perverse, would subsequently create many problems.

THE MATHEMATICIANS INVENT

SINCE ANCIENT TIMES, mathematics, in daily life, has been the servant of science and technology. Mathematicians were asked to solve concrete problems. Astronomical calculations, the prediction of eclipses, and land surveying provided the incentive and the justification for developing their ingenious techniques. This was the age of what we now call "applied mathematics."

Then, in the last century, the situation changed. Mathematicians began to contest their inferior status. They "freed" themselves. They invented their own problems and played gratuitous games whose rules they made up themselves. They behaved like card players who, while never cheating, constantly tinkered with the conventions of their favorite activity.

All this led to an astonishing discovery. The questions posed by science and technology represented only a tiny fraction of the problems that could be formulated. Most theories devised by mathematicians and published in specialized journals *had no application in reality*. In no way did they describe the world around us. Their axioms did not correspond to nature as we know it. They existed only for the pleasure they gave to the mathematicians who invented them. Their sole justification was their own internal coherence.

Mathematics thus took on a new dimension. It became a game for the human mind.

In current usage, "to invent" means to produce something that did not exist before.[1] We cannot say that Beethoven's *Ninth Symphony* "existed" before Ludwig sat down to his worktable. In the same way, a mathematician who, in playing with axioms, develops an infinite variety of non-Euclidian geometries is, like the artist, "inventing" or "creating."[2]

Under these circumstances, can we still take seriously Plato's and Descartes's idea that all of mathematics is inscribed in human memory? It would be about as believable as to affirm that Joseph Haydn had read, on scores already inscribed in his brain, the music of his 104 symphonies . . .

EINSTEIN AND GEOMETRY

AND YET . . . WE NOW HAVE to introduce a new element that will confuse the issue and revive the discussion. In the past, as I've already mentioned, mathematical theories were always created *in response* to technical problems. For example, at the beginning of the nineteenth century, Charles Fourier tried to explain the behavior of heat. In describing the increase in temperature of a bar of metal heated to its limit, he produced a new chapter in mathematics, the *Fourier series*, well known to all contemporary scientists.

Recently, the situation has many times been reversed. Mathematics has often anticipated physics. Here are a few typical examples.

At the beginning of this century, physics went through a series of major crises. Newton's theory of planetary orbits ran into difficulties. Mercury refused to obey, exactly, the dictates of his equations. The orbit was not a perfect ellipse. The discrepancy was very small, but as in Kepler's time, this tiny variation was to provide a rich store of information. It was Einstein who saved the situation, by reformulating the problem of the force of gravity. He discovered that the space in which planets move cannot adequately be described by the Euclidean geometry of our schoolrooms (two parallel lines will never meet . . .). To explain the subtle relationships between the bodies' masses and their movements, special geometries would have to be brought into play.

Einstein's colleagues informed him that such geometries already existed in the literature. They had been invented a century earlier by Gauss and Riemann. These two German mathematicians never imagined that their work would one day find a niche in the world of the stars.

Applying these new geometries — called non-Euclidean — to his problem, Einstein in 1915 developed the General Theory of Relativity. This rare gem of human thought does not content itself with correctly describing Mercury's orbit. It gives us access to the strange world of super-dense stars, as well as to the first moments of the universe.

HEISENBERG AND ALGEBRA

ATOMIC PHYSICS WAS also in crisis at the beginning of the century. There was no longer any correspondence between theoretical projections and laboratory results. This time, it was quantum physics that would save the situation. Bohr, Schrodinger, and Heisenberg were led to develop an atomic mechanics based on entirely new premises. What was missing was an appropriate mathematical theory. Fortunately, as in the case of Einstein and Gauss, the mathematicians had already passed that way. Heisenberg learned of a formalist mathematics called *linear algebra*, created a few years earlier by the mathematician Jordan. Adapted in such a way as to describe particles of matter, this formalism provided the basis for quantum theory. It was a dazzling success. In certain instances, this theory allowed one to predict experimental results with an accuracy greater than one part per billion.

There was a similar development in particle physics. The various natural forces — electromagnetic, gravitational, nuclear, and weak — although apparently very different, are, in some ways, alike. They can be described using analogue mathematical formalisms physicists call "gauge theories." Later, they realized that the French mathematician Cartan had already developed such a formalism under the name of "bundle theory." Introduce physical particles and forces and it works perfectly well . . .

TO INVENT, YES, BUT . . .

IN THESE THREE CASES we have seen how mathematical theories, invented in a spirit of play, have subsequently been used to describe the real world. We have seen how their structure and their internal logic have corresponded almost miraculously to the behavior of matter in certain specific circumstances.

The question arises once more: "How did mathematics come to humanity?" We are now faced with a paradox. If it is an "invention," how do we account for those few theories that adapt so well, and with such staggering effectiveness, when applied to nature? How is it that axioms invented by mathematicians totally ignorant of the problems to which they'll be applied succeed in replicating the subtleties of physical phenomena spanning over nine orders of magnitude?

On the other hand, leaving aside certain glaring exceptions, how do we justify the use of the word "discover" when almost all published mathematical theories result from the free play of mathematicians' minds and have nothing whatsoever to do with reality?[3] It would appear that the terms "discover" and "invent" are equally inadequate. To describe the mechanism that brought mathematics to humanity, we must look elsewhere.

MATHEMATICS IS NOT PERFECT

THE ELEMENT OF PLAY in mathematics modified somewhat the austere purity of the Pythagorean myth. New developments would go much further in this direction.

At the beginning of the twentieth century, mathematicians tried to strengthen the foundations of mathematics. They analyzed logic, the legacy of the Greek philosophers, paying particular attention to Aristotle. To their surprise, they uncovered a number of problems at the very core of the conceptual structure. Their attempts to correct the situation revealed other difficulties, more serious still.

Several schools of thought were founded in an effort to resolve them. They evolved in different directions and eventually failed to agree on even the most basic principles. No one knows if these conflicts will ever be cleared up. In his book *Mathematics, The Loss of Certainty*, Morris Kline describes the ups and downs of this turbulent chapter in the history of mathematics.

Here let us return to the testimony of Bertrand Russell, who at the start of his career sang the praises of mathematical perfection with such enthusiasm. In his autobiography, he describes his disappointment:

> I wanted certainty in the kind of way in which people want religious faith. But I discovered that many mathematical demonstrations, which my teachers expected me to accept, were full of fallacies, and that, if certainty were indeed discoverable in mathematics, it would be in a new field of mathematics, with more solid foundations than those that had hitherto been thought secure. But as the work proceeded, I was continually reminded of the fable about the elephant and the tortoise. Having constructed an elephant upon which the mathematical world could rest, I found the elephant tottering, and

> proceeded to construct a tortoise to keep the elephant from falling. But the tortoise was no more secure than the elephant, and after some twenty years of very arduous toil, I came to the conclusion that there was nothing more that I could do in the way of making mathematical knowledge indubitable.[4]

The myth of mathematical perfection was undone by these difficulties. Today, it is hard to believe that numerical laws might constitute a foundation for truth in the world. How could the core of reality, the language of God, be so riddled with dissonance?

We must concede, instead, that the development of mathematics reflects the workings and the limits of the human mind, of which, without a doubt, it is the product. But again, the question arises, more intriguing than ever, of its remarkable capacity to describe certain aspects of the real world.

THE MIRAGE OF A DEFINITIVE PHYSICS

ALONG WITH THE DREAM of a perfect mathematics, physicists for a long time nurtured the hope for a *comprehensive theory* of the universe. What Einstein called his "unified theory" reappears today with the ambitious label "theory of everything." In a typical burst of optimism, Stephen Hawking hints that it could very well see the light of day before the turn of the century.[5]

Where do we stand in this pursuit? Here we must call attention to an important fact: mathematics gives only an approximate representation of reality. Some historical background will help us assess the situation.

Among all the possible geometries, that of Euclid, which we were taught in our schoolrooms, continues to occupy a privileged position. The notions of angle and surface, with reasonings predicated on angles being equal, are used by thousands of architects and engineers throughout the world. The stability of bridges and skyscrapers testifies to their effectiveness in representing *adequately* the texture of reality. However, a geometrical surveyor with an excellent theodolite could demonstrate that these architectural renderings are not perfect.

In astronomy, Mercury's orbit serves as a comparable example. Euclid's geometry provides a serviceable description of the solar system, as long as the planets' positions are not calculated too precisely. Beyond a certain threshold, however, discrepancies between theory and observation become obvious.

Today, scientists admit that all theories of contemporary physics are approximate. None gives a universally valid picture of the real world. In addition, each operates within a given field, defined by certain conditions. Outside its own territory, it becomes useless.

For example, Newton's physics works perfectly well if the bodies to which it is applied are not moving very fast. When we want to describe the movement of particles approaching the speed of light, then everything breaks down. We must instead turn to Einstein's relativity theory. Perfectly suited to the study of high speeds, and also (in the General Theory) to the description of strong gravitational fields, it is unable, on the other hand, adequately to deal with the behavior of atoms. For this, we must resort to quantum physics. Its achievements in its own field are remarkable. It describes atomic and molecular phenomena with wonderful precision, *provided* these objects are not subjected to a strong gravitational field. In which case all bets are off . . .

To obtain valid results, we must first correctly identify our objectives and be prepared to settle for an incomplete description of nature.

"TERRA INCOGNITA"

WHEN EXPLORERS OF past centuries drew up maps of their voyages, they placed at the edge of the territories they had crossed a blank zone called "terra incognita"; these were the unknown lands about which they had nothing to say. This region's frontier separated the zone they had explored from still-virgin territory. It drew the line between the known and the unknown.

In physics, we have our own "terra incognita," a territory where nothing works. It's called, in technical literature, "Planck's conditions," and it is found at very high temperatures, around 10^{32} K, or at densities of 10^{93} grams per cubic centimeter. One might think that densities and temperatures this extreme play no role in our universe and can easily be ignored. Such is not the case. Comparable conditions very possibly existed in the earth's distant past, and the traces they left behind may have had a decisive influence on the fate of the universe. Unfortunately, our ignorance of the physical laws then in force prevents us from pushing our exploration of the world that far.

Many physicists are trying to formulate a theory appropriate to such extreme conditions.[6] The difficulties are enormous, and the results, to date, are disappointing. It would appear, at best, that we'll be able to push back our present boundaries only a little.

To sum up, we have become aware of the fact that mathematics can give us only *limited and approximate representations* of reality, even if these representations are sometimes extremely faithful. Their limits are known to us (though we often find it difficult to pin them down). Once we are beyond the range of their validity, physical theories lose all their value.

Year by year, the situation improves. Knowledge increases. Ignorance is in retreat. Will we, one day, be able to formulate a global and definitive theory of the universe? Or is this a mirage that recedes as fast as we approach it? Many researchers have dealt with this question, but there is no agreement on the answers. They are in large part a reflection, in my opinion, of the authors' own temperaments.

WHERE DO WE STAND?

Before continuing this inquiry, it would be a good idea to see how far we have come. The ancient vision of a perfect mathematics, set forth in the first chapter of this book, had, over the last two centuries, to contend with unforeseeable developments. It suffered a number of shocks. The first took place in the nineteenth century, when mathematicians invented a new game. It amused them to formulate brand new groupings of axioms, and to work out their consequences. They created from scratch a *multitude* of geometries, all different from that of Euclid. For the most part, these geometries described no reality and resolved no problem, but they did explore the element of play in mathematical activity.

Those (rare) mathematics that describe the real world raise different sorts of questions. Why them and not others? Why do they always apply only to a particular area, outside of which they are not effective? And why, within their own field, do they have this extraordinary capacity to describe what is real? The accuracy of their predictions — often to the billionth part — goes far beyond what, in our innocence, we would have foreseen.

A second blow to the bucolic simplicity of the Pythagorean vision took place at the beginning of the twentieth century, when a number of problems came to light regarding the internal coherence of logic itself. The myth of a geometer God's perfect language was seriously compromised. What then remains of the Pythagorean mythology and the prestige of mathematics?

Physicists, for their part, are forever seeking a definitive theory of the universe. Whether it will ever be achieved remains an open question. The inner contradictions of modern mathematics do not encourage much optimism. It's hard to imagine a definitive physics resting on the foundation of flawed logic . . .

IDEAS HAVE A HISTORY

CERTAIN SCIENTIFIC DISCOVERIES have profoundly altered our view of the world, but the shift in consciousness is often slow. New knowledge must first assert itself. Its influence must then filter down through the various layers of contemporary thought. That can take decades.

With some exceptions, the philosophers of antiquity thought the universe was *eternal and unchanging*. Matter always existed and would exist forever. Wood might decay and metal rust, of course, but these were "accidents," superficial transformations of no overall significance. On a celestial scale, all would be forever the same, and nothing would ever change.

This vision of a static universe pervades all of science, from ancient Greece to the beginning of the twentieth century. Neither Galileo, Descartes, Newton, nor Laplace called it into question. Einstein himself believed in it. In 1915, ten years before the landmark measurements of Hubble, Einstein's equations were already describing an evolving universe, but he went so far as to alter his own theory in order to preserve his conception of a static cosmos. When the movements of galaxies and the expansion of the universe were discovered, he recognized his error. "It's the biggest blunder of my life," he later said.

For several decades, various sciences, astronomy in particular, have presented the picture of a dynamic universe. *Ours is a changing world*. Not only do animal forms change, not only do stars evolve, change color, live, and die, but the overall properties of the cosmos — temperature, density, states of matter — alter significantly over the ages.

We live in *an evolving universe*. This discovery has influenced the way we think about everything, including thought itself. It affects our regard for the empire of numbers. It calls into question the Pythagorean mythology.

Humankind has not always existed. On astronomy's time scale, it appeared on the scene only recently. It emerged from a long series of ancestors, including primitive cells, metazoa, fish, amphibians, reptiles, mammals, and primates. Our ancestors' brains two million years ago were on average one-third the size of our own.

Humans have not always existed, and the Greeks were unaware of that fact. It's a reality that casts doubt on both the Pythagorean mythology and Auguste Comte's ladder of knowledge. For as we inquire into the nature of ideas and scientific knowledge, we must now introduce a historical dimension.

In so doing, we are faced with a fundamental and irrevocable fact. Logic and mathematics are creations of the human brain, but this brain appeared only after billions of years of evolution in a universe — ruled by physical laws — in which *thinking brains did not exist*.

Now here is the tricky question: *does mathematics "exist" when there is no brain to think it*? What is the meaning of the word "exist" when we're talking about ideas and there is no one there? Where, in the Precambrian era, were the figures and physical laws to which we owe our own existence hiding?

"Did two and two make four in the time of the dinosaurs?" asked Robert Pirsig, pertinently, in his beautiful book *Zen and the Art of Motorcycle Maintenance*.[7]

What was, at this period, the status of the empire of numbers? In the passage cited earlier, Descartes juxtaposes the assertion "I find within me countless ideas of things *which cannot be called nothing*" with the words "even though *they may not exist anywhere outside me*." It would be tempting to ask him our question: what does it mean for an idea or a number to "exist" when there is no one to think it? Or again: did two and two make four in the age of the dinosaurs?

THE NUMBER AND THE NUMBERER

THIS PUZZLE, IN FACT, is not new. Long before modern scientific discoveries, Aristotle had a glimpse of it and articulated its first "stage." In adding a second stage, modern astronomy brings us face to face with the true dimensions of the problem.

Aristotle questions the relationship between the existence of *time* and the existence of *a human psyche capable of creative thinking*. Astutely, he links this problem to the relationship between a *number* and the existence of *a "numberer" who thinks the number*.

"We may doubt," he writes, "that time can exist apart from the psyche. In fact, if we do not accept the existence of the *numberer*, we cannot envisage a *number*. But if it's true that, in the nature of things, only the psyche — or the intelligence within the psyche — can number, the existence of time without that of the psyche is impossible."[8]

In other words, Aristotle tells us that if there is no *numberer*, there is no *number*. This is the first stage. But what he doesn't know, and what we've since learned, is that the evolution of matter — from the Big Bang to the appearance of the human brain — obeys physical laws that the human brain expresses in numbers. The second stage thus states that if there is no *number*, there is no *numberer*. Which leads us inexorably to the following paradoxical conclusion: we need numbers to produce a numberer, and we need a numberer to dream up numbers . . .

A CAROUSEL FROM HELL

IN COLLOQUIAL LANGUAGE, we call this a "vicious circle." But not all vicious circles are equal. This one, as we will soon see, tops them all . . . To vice, it adds perversion. It has a way of reining us in just when we think we're finally out of its grasp . . .

Before trying to escape from our vicious circle, I'd like to describe its features in a little more detail. We can do so in several ways: for example, by setting side by side two apparently innocuous statements.

The first goes as follows: *the human brain is "inscribed" in laws*. I use the word "inscribe" in a very general sense. My purpose is to suggest two different things:

— on the one hand, that *cerebral activity* implies a series of physico-chemical reactions within the brain. These events require the interaction of a great many giant molecules. The behavior of these molecules is determined by electromagnetism and by the laws that govern this force;

— on the other hand, the *very existence* of the human brain, as a product of cosmic evolution, is linked to developments that stretch over fifteen billion years. These developments may be traced to the four forces of physics (nuclear, electromagnetic, weak, and gravitational) and the physical laws that govern them.

In other words, the growth in cosmic complexity over the ages, up to the emergence of thought, may be considered the realization of a potential that, *thanks to the existence of laws*, was already inherent in the primal magma.

Now we turn to the second statement: *laws are inscribed in the human brain*. This proposition reminds us of the fact that mathematics and physics are the relatively recent products of human beings.

These two assertions, side by side, are another version of the problem of the number and the numberer. Here is a third, more succinct: laws create the brain, and the brain creates laws. Each of these formulations, in its own way, plunges us into a vortex from which we will escape only with the greatest of difficulty.[9]

THE CIRCLE OF KNOWLEDGE

THE SWISS PSYCHOLOGIST Jean Piaget was one of the first to introduce a historical dimension into the study of the acquisition of knowledge. He recognized immediately that logic is a dynamic process, subject to evolution. In other words, the pedestal of pure bronze — the foundation for August Comte's ladder of sciences — is liable to the same questioning as the other disciplines perched upon the ladder. On what is it based, and from what does it derive its authority?

The question, once asked, gives rise to an irrefutable response: *the problem of the origin of logic is a problem psychological and biological in nature*.

"The universe," Piaget writes, "is known to man only through logic and mathematics, the product of his mind, but he can only know how he constructed mathematics and logic by studying himself psychologically and biologically, *that is, in terms of the entire universe*."[10]

This line of reasoning leads us to a new way of expressing our problem. Concepts are created by a brain whose activity may be studied scientifically in terms of physiology, or biochemical reactions. But this biochemistry is based on chemistry, which is based on physics, which is based on logic and mathematics, themselves a creation of the human brain. *Quite simply, we are back to square one . . .*

This fiasco leads us to substitute for August Comte's picture of a "ladder of knowledge" Piaget's image of a "circle of science." Taking as a starting point the series of disciplines resting one upon the other — psychology supported by chemistry, which is supported by physics, which is supported by mathematics and logic — Piaget proposes to close the circle by seeking the origin of these last disciplines in psychology itself.

Like a snake biting its tail, the ladder is bent back on itself, forming a kind of closed chain in which each science is a link. In its circular movement, the bronze pedestal is transformed before our eyes into a link in the chain.

This metamorphosis is crucial to our argument.

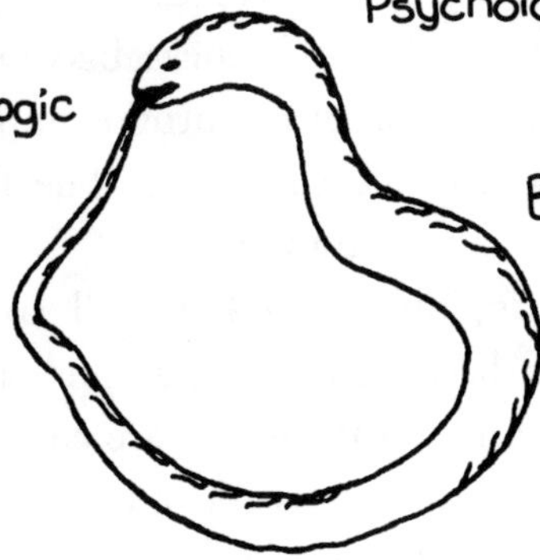

The serpent of science, Jean Piaget, about 1960.

THE CHICKEN AND THE EGG

THIS SITUATION BRINGS to mind the riddle of the chicken and the egg. Which came first? You need a chicken to make an egg; you need an egg to make a chicken. We're stuck. We're going in circles: egg, chicken, egg, chicken, etc. How to escape?

Today we've resolved the problem. Let's pay a visit to the henhouse. The solution to the chicken-egg dilemma might help us redirect our investigation.

To resolve the chicken-egg riddle, we need another ingredient: time. It takes millions of years of biological evolution for time to play its subtle but crucial role. It's clear where we must look: to the ancestors.

Let's retrace the unbroken line of chickens and eggs that culminates in our modern barnyard. Careful examination will enable us to detect gradual changes in our winged creatures. Very slight at first, the differences become significant and even decisive as our backward progression takes us several hundred million years into the past. The fowl become reptiles, then amphibians, and then fish.

But reptiles, amphibians, and fish all lay eggs that hatch and in their turn become reptiles, amphibians, and fish. Our riddle is still the same. And yet, we are close to an answer.

The existence of the egg is related to the phenomenon of sexual reproduction. The egg is where the two partners deposit their genes. Which raises the question: when did sexual reproduction make its appearance in the animal world?

In *Microcosmos: Four Billion Years of Evolution from our Microbial Ancestors*, by Margulis and Sagan, I've found a wealth of fascinating information on the sexuality of primitive cells.[11] The first living cells reproduced through division, without sexual contact, without laying eggs. Later, more complex cells invented sexual reproduction. Two partners deposited their genes in the same basket. This basket was called an egg.

Even if we don't know exactly how sexuality came about, we know that at a certain point a distant ancestor of the chicken perfected the technique of sexual union. This is what frees us from the vicious circle. Although this ancestor was neither an egg nor a chicken, it would give rise, after hundreds of millions of years, to the egg *and* the chicken.

We have solved the problem: what came first was neither the egg, nor the chicken, but this common ancestor that was neither egg nor chicken. We understand *how* we escaped from the vicious circle when we realize that it was not really a circle. Chickens out of the past resemble but *are not identical to* modern chickens. As we move back in time, the differences, at first tiny, become enormous and ultimately provide us with the key to the mystery.

Thanks to the historical dimension of evolution, we have solved a problem that seemed insoluble. This success suggests a possible approach. Applying the same key to the problem of the number and the numberer, we can ask a number of questions. What is the common ancestor, the shared origin of the number and the numberer? How did the two become different in the course of time? How did they evolve together, to assume their present form?

What form did number take when the numberer didn't exist? Don't forget that in its primitive form — whatever it was — number has been "operational" since the ancient days of the universe. Two hundred million years ago, Earth, populated by illiterate dinosaurs, was calmly orbiting the sun. It obeyed, unquestioning, "that" which was to become Newton's laws.

The lessons we learned in the henhouse suggest that the connection between thought and cosmos might lend itself to historical analysis. Let's try to establish a chronology. If we date the appearance on the scene of objects that are the study of different sciences, this is what we get.

The universe fifteen billion years ago was very different from the universe today. We can compare the state of matter at that time to a homogeneous soup of elementary particles. This ancient matter and the forces that acted on it may be studied using the traditional methods of *physics*.

In this primitive soup there were neither atoms nor molecules. Hydrogen would only appear a million years later.

Molecular chemistry requires the emergence of the first heavy atoms, and to manufacture these atoms, we must first have stars. Then the lives of these stars must run their course. When they die, they discharge into space the whole range of heavy nuclei they have spawned in their fiery cores. From these nuclei will come the first heavy atoms, and, when they combine, the first complex molecules. At the very least, we must allow *several hundred million years* to pass before cosmic *chemistry* can embark on its active phase.

Solid planets are much more recent. There must first exist, in sufficient quantities, the constituent chemicals for their rocky underpinnings: oxygen, silicon, iron, magnesium. This phase requires the combined activity of many generations of stars and takes *several billion years*. These planetary cores seem to be essential to the development of life. In the same way, the presence of a liquid aquatic layer seems indispensable to the play of prebiotic experimentation, raw material for *biochemistry*.

Where life is concerned, the stuff of *biology*, our observations are limited to our own planet. Here, the oldest fossils date from three billion five hundred million years ago. By this time the universe was already *at least ten billion years old*.

As for the first multicellular organisms (sea worms, medusae), we know of no remnants older than a billion years (counting from the present time). Over the ages, animals evolved and became more complex. The nervous system developed. Its commands were centralized in a brain. At first rudimentary, then more and more sophisticated, this organ dealt with information from the outside world. It processed it, organized it, and created the images from which it was able to picture for itself the surrounding world.

The human brain, which appeared *less than three million years ago*, pushed even further this effort to create an inner reflection of the outside world. It became the subject matter of *psychology*. This discipline tries to understand how the brain invents the concepts and numbers that enable it, among other things, to study the behavior of elementary particles in the primeval soup: a task for *physics*.

In a flash, we're thrust back *fifteen billion years*. Once again we find ourselves on an endless carousel, and we meet the snake that is chewing its tail. Our historical analysis has plunged us right back into the circle . . .

The circle of knowledge is far more vicious than that of the chicken and the egg. Its favorite device is to tempt us with false exits that, before we know what's going on, lure us back into its clutches, just when we thought we were home free.

THE CHALLENGE

THE TALE IS TOLD that in the twelfth century there lived three brothers called the princes of Serendip. It is said that these brothers knew how to turn the most adverse circumstances to their advantage. The word "serendipity" was coined to describe the art of profiting from the problems in one's way. We'll try to follow the princes' example, and we will soon realize that the difficulties in which we find ourselves can, in fact, be our guides. They'll point our inquiry in new directions. But first, let's briefly review the path we have taken.

By what right, we asked, do logic and numbers claim to account for all of reality? On what basis can they affirm that, wherever they have passed, there is nothing more to add, thereby stripping poetic expression of all but its own futility?

Auguste Comte's ladder seemed to support this claim, when logic was assumed to be the ground and final reference point for all knowledge. Then, in the course of our wanderings, we came upon a highly significant spectacle. The ladder of science was transformed into a "snake of knowledge." How does this spectacle help us to light our path? How does it address the question of science and poetry?

This metamorphosis represents yet another blow to the supremacist pretensions of scientific language. The snake of knowledge challenges the "fundamentality" of logic and Pythagorean ideas. These ideas have their own origins, which may be studied in terms of other sciences.

The circle of sciences shows us the limits of the scientific approach. However effective these disciplines may be within those limits, they are dependent on one another. Yet they operate in isolation. How, now, can they claim to monopolize reality and render null and void any other approach to the world?

3

Construction

THE HISTORY OF THE EMPIRE OF NUMBERS clearly demonstrates that logic is not something set down, final and unchanging, like those "dreams of stone" so dear to Baudelaire. Bertrand Russell has spoken of his inner conflicts and his hopeless quest for coherence. Logic must be seen as a process in flux, always perfectible, always open to question. Like the universe, logic is continually evolving.

"How did numbers come to humanity?" — the question torments us and takes on new connotations every time around. What happens to it when we introduce this fundamental notion: *the number and the numberer are subject to change*?

Jean Piaget proposes the word "to construct." This word implies both "discover" and "invent," but it goes much further. It posits the existence of building blocks, the identification of which will be our goal in the next chapter. Let's admit right off the bat: we won't be successful. But our failure will put us on another trail, one that's richer and more promising.

This idea of "construction" is not really new. Already, at the end of the eighteenth century, Kant was criticizing the Cartesian idea according to which we "discover" mathematics in our memory. He emphasized the active role of the human brain in the development of knowledge. The brain "constructs," but out of what?

Kant responds to this question with an authoritarian dictum. There are "*a priori* givens," including *time* and *space*, as well as "categories" of understanding, such as *causality*. These givens, pre-existent in thought, serve to organize our sense perceptions. This is the origin of scientific knowledge.

Here, we encounter once more our earlier questions. Do *time*, *space*, and *causality* exist when there is no brain to think them? Kant, it must be said, simply avoids the problem. Moreover, developments in contemporary physics seriously call into question the idea that time and space can be "*a priori* givens."[1]

Let's take the question a little further. Where and when did logic, as a discipline, first emerge? What is its history? What do we know of its development, as it relates to both biological evolution and the individual? When, in global history, did reasoning first make its appearance?

Many scientific disciplines, to one degree or another, have taken an interest in the origins of logic. Four sets of researchers have approached the problem from different angles. The *ethologists* have studied intelligence in the animal world. The *anthropologists* have tried to reconstruct the workings of the primitive human psyche. The *psychologists* have studied the burgeoning of thought in young children. The *neurophysiologists* have tried to reach an understanding of what goes on in an active brain.

In the following pages we'll interrogate these researchers. Essentially, they'll show us that logic, in however rudimentary a form, is much older than we have been willing to believe up to now. They'll confirm that, in primitive man as in the child, logic is "constructed" from prior elements that we are not able adequately to pin down.

We'll see how some researchers have been tempted to identify these elements with the molecular structure of the brain. A little reflection will reveal the hidden pitfalls behind such an identification. It lands us back in the vicious circle of knowledge. Once again, we'll discover that our search is not yet at an end.

CAN BIRDS COUNT?

NOTHING IS MORE INSTRUCTIVE for those who want to explore the origins of the human psyche than ethological research. The study of the lives of animals leads to astonishing revelations. From it, we can gather extremely pertinent data. In more or less rudimentary form, we encounter our own activities, behavior, and shortcomings . . .

Did we really invent the art of manipulating numbers? Who, on our planet, was the first to carry out mathematical operations? It would seem that certain small "bird brains" were there long before us.

Before venturing into a granary, crows make sure there is no one inside. If three people go in, *together or one at a time*, the bird outside will wait for the three people to leave, *together or one at a time*. We can repeat the exercise with four or five people and the result is the same. Certain crows "strong in math" can go as high as six.

Can we say they know how to "count" up to six? If a child were doing the same thing, we would take it for granted.

Some specialists contest this conclusion and characterize it as "simplistic." In any case, we have to admit that, one way or another, these birds are familiar with numbers. They know how to use them. But what rudimentary form do they assume in their brains? As with the chicken and the egg, we may ask what precursor evolved to the point of taking on, in our heads, the shape of a number? What is the driving force behind this conceptual evolution?[2]

These experiments, and many others, show us that the "embryos" of number and reasoning were already present in our animal ancestors. They form part of a heritage that, in some mysterious way, has been transmitted across the ages.

Our own thought processes and logic have emerged from these embryonic elements. We've neither invented nor discovered mathematics. It's grown out of mysterious "primitive givens" inherited from our animal ancestry.

THE "EMPTY BOXES"

WHEN AND HOW DID LOGIC first appear among prehistoric man? How did the Lascaux painters, for example, think? How did they reason? Here we can capitalize on the fact that certain communities today lead an existence similar to that of Cro-Magnon man. May we assume that similarities in behavior reflect a similarity in mental attitudes? That the mind of an Australian aboriginal works like that of our distant ancestors? That to study what is considered a "primitive mentality" will cast light on our past? Up to a point, doubtless. But what point?

Anthropologists have amassed an enormous fund of documentation regarding the behavior and way of life of tribal populations scattered across the planet. It's difficult for a nonspecialist to come to terms with it. I've taken a special interest in the monumental body of work by Claude Lévi-Strauss. Here is an author who is not afraid to provide, now and then, his own personal interpretations, extremely useful for the uninitiated.

Lévi-Strauss's primary contribution, in my opinion, is his intuition that simple and familiar aspects of human life, such as styles of nutrition or family relations, put us in direct contact with the workings of the mind.

Comparing different tribes, he brings to light some revealing common traits. The themes of family ties, dominance, and incest taboos may be found everywhere on the planet. They are universal in character, with their own local particularities. It's much the same with the tales and legends that oral transmission has preserved in memory. Despite the widely diverse features proper to each tribe, the structure and the themes of the stories show surprising similarities.

To the natural objection "But people communicate among themselves and tell each other stories," Lévi-Strauss replies, pertinently, that what is important is not to know if people invented — or simply borrowed — similar legends. What is important is that the legends endure and become part and parcel of each local tradition. The longevity of the great legendary themes shows how rooted they are in the deepest layers of the human spirit.

These resemblances are evidence, says Lévi-Strauss, for the existence of fundamental and universal structures in the mind. He speaks of a "structural unconscious" that influences and shapes the workings of the human psyche.[3] It's here that we find the "empty boxes" that tend to contain and mold human behavior.

It is reflections such as these that give rise to the idea of *something prior to* logical thought: a common heritage, transmitted from one generation to another.

And what does Lévi-Strauss have to say about the origin of these "empty boxes" that seem to play such an important role in the evolution of the human psyche? Very little, overall. In *The Naked Man*, he associates them with "that obscure drive which, over millions of years and with the aid of tortuous and complicated devices, has ensured the pollinization of orchids, thanks to transparent windows allowing the light through . . ."[4]

HOW CHILDREN LEARN TO COUNT

If, for lack of documentation, the historical origins of human logic are in large part beyond our reach, there is a field of study more easily accessible: the psychological evolution of children. How do little ones enter into the world of reasoning?

Here, I've immersed myself in the work of Jean Piaget. Along with his co-workers, Piaget observed and questioned a multitude of preschool infants. Through shrewd testing, he followed their mental development, from the earliest age until the beginning of school.

We present a small child with two glasses of different geometric shape. The first, slim as a flute glass, is brimming with water. The second, wide and squat, is empty. While the child watches, we pour the water from the first into the second. The latter is now also full to the brim. We ask: "Which glass contains the most water?"

Very small children don't know what to reply. The difference in form between the two glasses troubles them and prevents them from finding the answer. And then, one day, it comes. Just like that. Without hesitation, they will say, "It's the same." Intuitively, perhaps without being able to say how, they have understood. *The form is of no importance.*

All the water in the first glass is now in the second glass. The quantity of water has been conserved during the decantation. This conservation, associated with the fact that the two glasses were, one after the other, exactly filled, determines the response.

The notion of conservation is one of the linchpins of modern physics. It has been applied with equal success to a large number of different "substances." It controls the behavior of energy, many forms of motion, electric charges, as well as other charges associated with different natural forces.

A remarkable aspect of these pedagogical experiments is the "lightning stroke" that generally accompanies the discovery. The correct answer appears in a flash, often with an intense feeling of pleasure. What seemed up to then to pose an insurmountable problem (how to compare different forms?) becomes a self-evident "fact" (the forms are of no importance). In an instant, the new data is understood and assimilated. Permanently. It will nevermore be called into question.[5]

Piaget and many psychologists that followed him have emphasized the importance of the "external world," here represented by the glasses of water. Logic is a product of the interaction between the human intellect, on the one hand, and reality as perceived by the senses, on the other. Cut off from the world, without any contact with things or people, a child will not develop. Furthermore, we are now aware of the crucial importance of emotional factors in education. The good teacher is above all someone who, through a sympathetic relationship, inspires confidence in the child.

Descartes seems oblivious to all this. "It seems like noticing . . . things which were long present within me although I had never turned my mental gaze on them before," he says. Piaget would reply, with reason, that without the long and active relationship that he had maintained, since childhood, with the outside world, he would never have been able to "notice things within him."

THE "STARTING POINTS"

WHAT DOES ALL THIS TEACH US about the origins of logic? In a book of interviews with Piaget by Jean-Claude Bringuier, I found the following comment, very much in line with our own inquiry (the parentheses and italics are my own):

> I think that all structures (mental structures, or concepts and units of reasoning) *are constructed*, and that what is important is how that construction proceeds. Nothing is given at the outset, other than *a few starting points*, on which everything else depends. But the structures are not provided in advance by the human mind nor by the external world such as we perceive it or order it. They

are constructed by an interaction between the *activities of the subject* and the *reactions of the object*.[6]

Once again, we encounter the word "construct" — a word that relates both to "*discover*" (we construct out of something that already exists) and "*invent*" (but the construction did not exist as such). Piaget takes us up to a new frontier, well defined by the hermetic utterance "*nothing* is given at the outset, other than *a few starting points*, on which everything else depends." What is new, when we compare this to Kant, is the acknowledgment of our own ignorance relative to *that-out-of-which* the brain does its constructing. The words "a few starting points," like the phrases "empty boxes" and "obscure drive" of Lévi-Strauss, raise more questions than they answer . . .

THE BRAIN'S "MACHINERY"

ELSEWHERE, PIAGET HAS let slip some of his thoughts on these "starting points." He speaks of his hope that biochemical studies of the brain will be able to relate the starting points to molecular phenomena.

Similar opinions have been expressed by other psychologists and art critics. Freud, in his text *Leonardo da Vinci, A Study in Psychosexuality*, posits an organic origin for the "psychic structure." The educationalist Arno Stern, after having discerned similarities between children's paintings among the tribes of New Guinea and those of little Westerners, offered in explanation the hypothesis of an "organic memory" for forms.[7]

We know intuitively, and scientific research has confirmed, that it all takes place in the brain. Contained in its shell of bone, this organ is the seat of our emotions and our thoughts. What is the relationship between abstract reasoning and neuronal activity in that opaque mass between our ears?

Let's put the question to those researchers who, with a variety of different instruments, have tried to shed light on the workings of the brain.[8] In recent years, much progress has been made in this field.[9] Then, as is our practice, we'll reflect on the impact of this new knowledge. How can these results help us on our journey?

The brain has often been compared to a gigantic computer. It's a poor comparison at best, a fitting reflection of how little we really know. We'll make use of it for what it can teach us, while not losing sight of its limitations.

The brain is made up of hundreds of billions of cells called neurons. Through a network of connections, each neuron is linked to a large number of other neurons. The number of possible connections in a human brain is immense. This does, indeed, suggest the dense wiring of electronic circuits in computers.

Cerebral activity consists of the remarkable crackling of a host of neuronal discharges, impulses that travel through these connections, triggering detonations of specific molecules that send forth signals whose effects will in turn excite other neurons. Let's bear in mind, above all, the importance of electrical phenomena (the impulses emitted by neurons when they discharge) and chemical reactions (that make it possible for a signal to pass to another neuron).[10]

MENTAL IMAGES AND NEURONAL DISCHARGES

WHAT MIGHT BE THE RELATIONSHIP between these countless discharges and the reasoning process?

We're far, of course, from being able to answer this question. However, remarkable advances have been made in this regard. Reasoning is founded on concepts, concepts arise out of images, and images are the product of sensory perception. The work of two American researchers, Huber and Wiesel, has enabled us to link image perception to the individual behavior of neurons. This is already a big step in the right direction.

Studying the vision of cats, these researchers have shown that each component of an image is perceived by different groups of neurons. Some groups discharge when a horizontal line is present in the landscape. Others react only to vertical lines. Still others to certain oblique lines exclusively.

Some neurons do not react if the image is static. They react only to movement. There are those that are sensitive to rapid movement; others to movement that is slow. Various directions — bottom to top, left to right — are associated with their own groups of neurons.

It all happens as though the landscape were being *analyzed* by the eye. It is decomposed into different elements: lines, colors, movements. Perceived by different groups of neurons, these elements — akin to letters of the alphabet that make up a language — are then reconstituted in the cerebral cortex.

MOLECULES AND MEMORY

IMAGES FORMED IN THIS WAY are recorded in the memory, where they are then made available to the associative brain.

How are these images stored? What happens within your cranium when, on being asked, you provide your telephone number or name the capital of Poland?

Many researchers have considered this phenomenon. First hypothesis: the bits of information are linked to specific molecules. Our library of memories would be composed of a large number of atomic structures, duly labeled, ready to be recalled, as one would seek a document on the shelves. This idea, very popular a few years ago, has been tested in a number of ways. The results have generally been negative. Most researchers have discarded it, at least in its most simplistic form.

Other forms of information storage have been proposed. Perhaps memories activate connections between neurons.[11] The problem remains on the agenda.

THE MIND OF THE MOLECULE AND MOLECULES OF THE MIND

THE PHARMACOLOGY OF MENTAL ILLNESS enables us to explore other relationships between psychic activity and molecules. Schizophrenic behavior, for instance, is associated with an excessive quantity of a specific molecule called *dopamine*. We relieve the symptoms by injecting substances that neutralize the effects of dopamine. In the same way, premature senility (Alzheimer's disease) would seem to be related to a lack of *acetylcholine*.

Thus, image perception, storage in memory, and certain aspects of mental activity can be associated with electrical and chemical phenomena within the brain. These neurophysiological discoveries support Piaget's hunch: biochemical research into the brain could link the famous *starting points* to molecular phenomena.

These spectacular advances have given rise to the idea that molecules could hold the secret and provide the final explanation for all human behavior, including the origins of thought, language, artistic activity, and metaphysical angst. They would be in some sense the primal mechanism of consciousness, the material for Piaget's "starting points" and Lévi-Strauss's "empty boxes."[12] In France, Jean-Pierre Changeux has emerged as the prophet for this molecular vision of reality. "Man has no need of mind," he proclaims triumphantly in his book *Neuronal Man*.[13]

A MOLECULAR MYTHOLOGY

OUR ANALYSIS OF THE CIRCLE OF KNOWLEDGE has shown us the inadequacy of such an ideology. Far from getting us off our infernal carousel, it lands us right back on again. For if we explain numbers, laws, and thought in terms of molecules, we must, in return, express the physical chemistry of molecules in terms of numbers, laws, and therefore thought. We are immediately faced, once again, with the problem of existing laws and their relationship to human thought. *If mind comes from molecules, molecules come from the mind.* To Changeux we would reply that, without the "mind" that he rejects, he could not speak of the "molecules" that he invokes. Nor could he express his proposition: "Man has no need of mind."

We now see that, in a spectacular about-turn relative to the *Pythagorean mythology*, recent scientific discoveries have given rise to a *molecular mythology*. Both these mythologies, each in its own way, are mirages, facile solutions.[14] They are temptations to be avoided by those who want to explore what lies behind our capacity to understand the world. Like Ulysses, we must seek to navigate between the Charybdis of "primal ideas" and the Scylla of "all-explaining molecules."

Before concluding this section on the mechanics of the brain, I must mention a particularly significant discovery, one whose implications are far from having been exhaustively explored. Our brain is divided into two distinct parts, approximately situated to the left and right, each associated with different mental operations. In general, the left brain analyzes while the right brain synthesizes. Words, concepts, and ideas prevail on the left, while the right proceeds by synthesis and takes a comprehensive view.

What's important here for us is the coexistence in our brain of *different ways* to represent the world. Our knowledge is neither limited, nor reducible to the sum of things we know and the number of mental operations we can perform. This discovery may be regarded as yet another challenge to the dominance of Pythagorean thought.

WHERE DO WE STAND?

BEFORE EMBARKING ON NEW PATHS, it would be best to pause and review briefly the ground we have covered in the first three chapters.

We began with the question: does scientific knowledge of the universe mean an end to poetry? Is nature's marvelous scenic beauty nothing more than the working out of solutions to mathematical equations in physics? Can we still respond to a sea on fire with the setting sun once we're acquainted with Maxwell's theory?

To question the legitimacy of these discouraging pretensions on the part of a certain scientific discourse, we journeyed back 2,500 years to the shores of the Ionian sea, where the scientific approach was born.

The discovery of the importance of mathematics in the workings of nature gave rise to what we have called the *Pythagorean mythology*. Numbers are the ultimate reality. Science derives from a numerical language (Auguste Comte's ladder). The world is a product of mathematics. The human brain discovered mathematics already inscribed inside itself.

Later on, this picture of innocence was to receive a number of blows from which it would never recover.

First came the element of play in a scientific world. A profusion of mathematics were invented that had no "use." Only a few described the real world.

A second source of concern: the internal contradictions. Nothing led one to believe that some day mathematics would resolve its problems and be whole. Given its weaknesses and its incompleteness, its amazing effectiveness seemed even more of a mystery.

The third challenge came from astronomy. Nature was working well, long before mathematics was ever devised. By asking where concepts came from, Piaget toppled August Comte's ladder. The source of logic is a scientific problem. The sciences all grow out of one another.

To find a way out of the circle of sciences, we tried to retrace the history of thought. The work of ethologists showed us that we were not the first to think. Lévi-Strauss spoke to us of "primitive structures" in the human mind, and Piaget of the "construction" of concepts and of reasoning from mysterious elements. We tried to associate these elements with molecular phenomena by studying the brain's biochemistry. But in realizing that it's the active brain that studies biochemistry, we saw that this approach just brought us back to the circle of sciences. In Chapter Four, we will seek a way out along the convoluted paths of psychoanalysis.

4

The Zones of Mental Construction

ON FURTHER REFLECTION, it was perhaps wrong to reproach Piaget and Lévi-Strauss for the vague terminology of their final analyses ("starting points," "empty boxes"). Let's look upon them rather as a sign of hope, the inklings of a way out of the circle of sciences. Would it not, indeed, be paradoxical if we could express, in clear concepts, a theory concerning the origin of concepts? Can one explain logic in terms of logic?

In this fourth chapter we'll be setting off in another direction, diametrically opposed, one might say, to our earlier path. We'll be addressing the least "logical" of all scientific disciplines: psychoanalysis.

The access psychoanalysis provides to something that's beyond the scope of the intellect, but which surrounds and abuts its territory, tips us off to a new approach. Rather than seeking (in vain) to understand *how* rational thought came to be, we'll ask ourselves *where* it came into existence. This new undertaking will plunge us into the murky waters of the unconscious. There we'll discover the common source of logic and of poetry . . .

Just as Christopher Columbus discovered America, Freud opened our eyes to an immense continent, swarming with primitive urges. Beneath a thin layer of consciousness, he encountered in the raw the primal instincts of animal life: sexual impulses, violence, and aggressiveness. Psychoanalysis, like the theory of Darwinian evolution, links us to our biological ancestry. It discerns, behind our "civilized" behavior, the weighty legacy bequeathed us by our ancestors. All of which makes for a life both rich and difficult.

For the purposes of our inquiry, we'll draw on Freud's assertion that the "said" has its source in the "unsaid," the verbal in the nonverbal. Behind all human activity, driving it and steering it, may be glimpsed a hidden reality: desire. Not a desire for one thing in particular — to eat, to make love (which disappears once sated) — but an unquenchable desire, at the roots of our emotions, in the very fabric of our bodies, in the furthest reaches of our human identity. It is, in fact, a desire for the impossible return to the mother, to that time of union when she and the child were one. It is into this dark world that we now must venture!

FULL BOXES

THE SWISS PSYCHOANALYST Carl Jung studied at length the problem of the origin of thought. According to him, there is a collective dimension to the unconscious, inhabited by universal "archetypes" that underlie the motivating forces at work in a human being. These archetypes have played a crucial role in our evolution.

For a novice reader, it would be tempting to identify Jung's "archetypes" with Lévi-Strauss's "empty boxes." Lévi-Strauss, however, is vehemently opposed to any such comparison.

The difference, fundamental for him, is that his "boxes" really are "empty," while Jung's are "full." That was his response when, a few years ago, taken aback by his sharp reaction to the very idea of such a resemblance, which to me seemed natural, I asked him for some clarification.[1]

Empty of what? Full of what? Herein lies the substantial new contribution that psychoanalysis can make to our inquiry. It focuses on the *affective charge* associated with mental activity. The psyche's basic elements are no longer regarded as formal structures but as active players, charged with emotion.

Jung's "archetypes" are noteworthy for the tremendous energy they can generate. We have seen, on our television screens, the hysteria of Iranian crowds on the death of Khomeini or the fanaticism of young Nazis, carrying torches and surrounded by huge swastikas, ready to die for a mad dream. These images illustrate the immense energy that may be mobilized by such archetypes as the father, the leader, holy war, the nation, the native land.

LOGIC AND THE BODY

TO MEASURE THE DISTANCE we've traveled since our inquiry began, let us remember that Descartes wanted, at all costs, to separate body and mind. We are now moving in the opposite direction. By emphasizing the emotional environment that envelops the primal flux of thought and logic, psychoanalysis demonstrates to what degree these realities are rooted in feeling, and therefore in the body itself.

Here, again, we could be seduced by the so-called "materialist" theory, according to which thought is a product of the body, just as bile is produced by the liver or your watch by the watchmaker. This thesis would in no time at all get us back into the very rut we're trying to avoid. To "explain" thought in terms of the body would be like grounding psychology in biochemistry, which, and here we go again, would derive from chemistry, which in turn . . . And we'd be cornered once more. If we want to elude the vicious circle of sciences, we must never let down our guard . . .

THE STOMACH TIGHTENS . . .

It's clear that the body that concerns us here is not just an idea. It is *that* which we experience every day, prompting Wilhelm Reich to say: "We do not *have* a body, we *are* a body." *That* thanks to which we are in this world and will remain here as long as it will allow. *That* which makes us feel, suffer, and experience anxiety . . .

A great teaching of Oriental wisdom, one that distinguishes it from Western idealism, is precisely the primacy of the real body. The harmonious functioning of a human being must involve a "reconciliation" with the body. That's the essential contribution of yoga, for example, which has been revived in our own day by so-called "Californian" therapies: bioenergetics, Gestalt, etc.

Psychoanalysis knows this, confronted as it is, day in and day out, with the suffering imposed by the split between body and mind. I'd like to cite here a beautiful text by the French psychoanalyst Elie Humbert. It's an expression of what he himself has lived and what his analysands have taught him.

The body exists and the mind doesn't know what that means. It can find no vantage point that will enable it to comprehend this existence, and so the mind ponders its significance . . .

It is the body that dies, and in so doing proves that it exists. It is the mind that questions, and finds itself drawn beyond its habitual ways of thinking. It no longer moves from one signifier to another, to grasp those connections that weave the real world together. The question is of a different order, because now it is linked to the body. It has become the experience of something not there, but that the entire body feels. It is no longer a matter of reflection, but of rumination and anguish.

The stomach tightens, the whole body suffers; it can become ill, and that we call metaphysical angst. It's a term we use when one of the mind's most exalted activities hollows out the belly and unsettles the organism.[2]

THE TERROR OF THE "NOT-ME"

WHAT DOES ALL THIS tell us about the origins of thought in human beings?

Rather than ask Freud — he studied childhood development only through adult neuroses — we'll question researchers such as Melanie Klein, D. W. Winnicott, and Françoise Dolto.[3] These psychologists were interested specifically in children. By concentrating on the affective side of the problem, they correctly pinpointed where it all begins.

The outside factor that triggers a child's mental activity is, quite simply, "reality," but in what one might term its *fearful dimension*. In order to live, we have learned to suppress the terrifying aspects of the real world. They come back to us sometimes and jump out at us in the form of suffering, death, and all kinds of horror. We have only to read the newspapers and look at certain television reports. They come to us out of Yugoslavia, Somalia, South Africa, etc. It's this side of reality that concerns us now.

Poets keep their eyes open. Listen to Michèle Lalonde, a poet from Quebec: "Reality cannot be reduced to language. It must be approached with a certain dread, mixed with pride, like matter in fusion and flight, existential magma, everlasting chaos."[4]

The crucial event in our life, says Winnicott, is our *encounter with reality*. This meeting is the source of all anguish, the central drama in human existence. To his mind, no human being can ever escape the tension created when inner and outer reality confront each other. The gradual acceptance of reality is a task that never ends.

FROM TEDDY BEAR TO TELESCOPE

AT BIRTH, THE CHILD is one with his mother. He lives in a cocoon that isolates him from the outside world. His gradual coming to terms with reality represents the story line and the drama of his existence.

Here Winnicott made an important discovery. The encounter between inner and outer reality takes place on *middle ground between the ego and the external world*. In this transitional territory, the child can *act* on reality, transform it, re-create it. It's through such activity that he can exorcise the element of fear.

The key word here is *play*. Winnicott talks of the playground where the small child is "lost" in play. With his teddy bear or rag doll, he finds relief from the anxiety caused by the external world and the tension created by the necessity of facing up to it.[5]

Winnicott directs our gaze toward that playground where, in response to the tension generated by an encounter with the world, we see the emergence of human invention and creativity. Nourished by this tension, they become the focal point for all psychic activities. The telescope or the paintbrush takes the place of the stuffed toy. In this garden religions, arts, and sciences all flourish, helter-skelter.

Each of these activities alters in its own way external reality and makes life livable. Each plays a specific role in this ambitious project of "reconstruction" that allows the human being successfully to face up to the outside world and live to the full his life on earth.

Some children never succeed in this, and close themselves off in their "autism." Why? Psychoanalysts lay much emphasis on the crucial role of the mother. By encouraging the development of this zone of play, she alone can make bearable the anguish and loneliness associated with separation.[6]

In short, psychoanalysis leads us to seek the origins of thought beyond thought itself. It indicates the site, the context, and the procedures involved in the beginnings of mental activity. There we have discovered the existence of an extremely fertile zone where the human being comes in contact with the outside world, a playground where the ego rebuilds its reality, where it can face up to the "terror of the not-me" and live fully. Creativity is for Winnicott the criterion of a successful life.

THE NAMES OF FLOWERS

IN THE COURSE OF OUR INQUIRY, we've garnered a bit of data that's particularly instructive. *It's in the self-same soil, nourished by childhood fears, that scientific, poetic, and religious activity is conceived.*

These three sisters, triplets at birth, have been given the same task: to remake the world. Throughout this book, up to now, we've been disputing the pretensions of the first to supplant the second. In the pages that follow, I'll try to show how their methods are both different and complementary.

By decoding nature's behavior, science has succeeded, up to a certain point, in casting out fright. It reassures. Thunder and comets no longer scare us. We are still helpless before the destructive violence of hurricanes, but we can give advance warning to populations that are threatened. Once-fatal illnesses can now be contained by medicine.

To reach its goal, science must scout for facts. This requires controlled experiments performed in the laboratory. The reality thus re-created — as theories and laws — must of necessity reproduce the observations. This constraint gives science its credibility, but it also limits its expressive powers. Words are used only to impart information.

Committed to rigor and precision, the scientist must express himself in clearly defined language, free of any ambiguity. He constructs his sentences with an impeccable logic that ensures the effective transmission of data.

A quasar is called "0957 + 51." This name is both precise and practical — it gives the position of the star by its celestial coordinates. On the other hand, it has no affective connotation. It does not lead us to dream . . . This is the price we pay for useful information.[7]

Poetry uses language to different ends. This Japanese haiku is an excellent example.

A blade of wild grass
When I knew its name
I found it more beautiful.

Beauty is an experience of the world. It includes both the outside world and the inner world of perception. It is situated in that middle ground of which psychoanalysts speak. In putting words to reality, poetry enriches our relationship with things. We see them *differently*. We are made, quite simply, to see them. Once named, the flower *becomes* more beautiful.

Sunsets can never be the same for those who know these lines of Baudelaire:

By late afternoon
the canals catch fire
as sunset glorifies the town;
the world turns to gold
as it falls asleep
in a fervent light.[8]

The fire, the gold, and the fervent light come together for the reader, and will continue to enhance those moments when the sun goes down.

THE "CLASH" OF WORDS

TO ATTAIN ITS ENDS, science is duty-bound to be cool in its use of language. Poetry uses it in a more dynamic, inventive way. Unlike the scientist, the poet adores ambiguous words, teeming with multiple meanings, full of connotations picked up through the ages. He juggles the order of words; he knocks them against each other, like stones, to throw off sparks. Turning ideas on their head, putting side by side in an unexpected way terms that don't go together, he invents images, impressions, strange feelings, a new way to experience the world.

In the space created by this disorientation, there is a sudden radiance, and we glimpse a new "meaning" that cannot be reduced to the words that gave rise to it. In contrast to scientific discourse, *the less a poem means* — in the usual sense of the word — *the more likely it is to have "meaning."* "The sea naive as a convolvulus," writes René Char. Reread this line and let yourself go . . .

Poetry is a different way of approaching the hidden flux of reality. It enhances our capacity for linguistic expression. Moving beyond the utilitarian, it finds fresh avenues for describing the world, for probing its unexplored riches. In a spirit of play, it creates newfound lands.

Scientific language is perfectly adapted to the *analysis* of strict facts. But, to give an overview, *to encompass all aspects of a subject*, poetic language is much more effective. I experienced this recently, first hand. Having completed my book on the history of the universe, for a long time I sought an appropriate title. My first choice, "cosmic evolution," was accurate, dull, and flat. I used it as my subtitle. For a title, I chose a line from Valéry, *Patience dans l'azur* (*Atoms of Silence* in the English edition), infinitely more evocative.

On the paths beyond thought, music goes even further. Its medium is sound, stripped of any conceptual meaning. Like logic, its structure is often rigorous and strict, but it too can create new emotions through its art.

THE ABYSS OF THE UNTHOUGHT

"Existential magma, everlasting chaos," writes Michèle Lalonde. "Reality cannot be reduced to language, is indifferent to Logos." I don't know if we have to follow the poet that far. Let's put the question differently: can we be sure there is no rapport between the true nature of things and what we can know and say about them? Is it correct to think, as she does, that "Any writer worthy of the name works with the illusion and the megalomaniacal intention of grasping reality, and in fact succeeds only in defining a relationship to it and taking a precise position"?

We may certainly have our doubts. For we would then have to despair of understanding why mathematics (some mathematics) is so effective in describing certain aspects of the real world.[9]

If, on the other hand, we seek the beginnings of ideas in a dimension inaccessible to the intellect alone, but *sensed* by the body and related to what psychoanalysts call the "unconscious," then we can no longer talk of indifference. What we say about things reveals something of their true nature, just as the visible part of an iceberg indicates the presence of an invisible mass lurking below.

Is this, however, an adequate explanation for the wonderful effectiveness of physical laws? Many writers have studied the question. Let us admit outright that no really satisfying answer has been offered up to now . . . We are left with vague intuitions, plausible but unprovable.

In a few words or in many, all these writers say more or less the same thing. It can be expressed in the following way. The human brain is one of the products of cosmic evolution. Its development has been shaped by the action of physical laws on matter in the universe. This accounts for its capacity to formulate the principles according to which, and thanks to which, it has been structured.[10]

In other words, to the sentence, *The universe is intelligible to us because we are intelligent*, we may add, *we are intelligent because the universe is intelligible.*

Such affirmations are unconvincing to rigorous minds. They are unprovable and "unfalsifiable," as the philosopher Karl Popper would say. Nevertheless it's possible that, in some sense, they correspond to reality.

The idea of intelligibility raises the question, once more, of order in the universe. Why is there not chaos? But what is chaos? How can we even talk of chaos? This question brings us face to face with the limitations of language. We can only talk of chaos once we've defined it. And we can only define it in terms of order, which brings us back to the world of the intelligible. "Chaos," like "nothingness," is beyond all intelligibility.[11]

What more can we say? We can only peer over the edge of the abyss of the "unthought," from the depths of which, mysteriously, both reason and poetry emerged.[12] The resulting vertigo influences how we feel about ideas. It guards us from the siren song of certitude and the tyranny of logical thought.

We understand better, from this vantage point, Jacques Lacan's words:

I always tell the truth,
not all,
because to tell all,
is more than we can do,
there are not enough words.
It's by this very
impossibility
that truth clings to what's real.

In other words: behind the capacity of words to grasp reality, we sense, continually in flux, layers impervious to understanding. Reason by itself is not able to do them justice. To clearly perceive the sun setting over the ocean, *all human faculties must be brought into play*. It is Schiller's message:

The full mind is alone the clear.

PART TWO

BUTTERFLIES OVER A FIELD OF COLZA

Science and Freedom

5

Butterflies and the Laws of Physics

Will new and alive the beautiful today
Shatter with a blow of drunken wing
This hard lake, forgotten, haunted under rime
By the transparent glacier, flights unflown!
"The Swan," Stéphane Mallarmé[1]

THE ACID YELLOW OF A FIELD OF COLZA lights up the spring. Above the blossoms two butterflies flutter, nonchalant. Now there are three, two yellow and onc white. They turn, twirl, and climb slowly into the sky. High up, the third flies off. I'm barely able to follow it with my eyes. It heads toward the sun and disappears into the glare of a white cloud.

"The world is all that is the case," writes the Viennese philosopher Wittgenstein. Reality is a tapestry of unique events that animate each day.

The "present moment" poses problems for scientific thinkers. For a long time they tried to deny "today." It was held down by a series of restraints. In the name of "determinism," it was denied its freshness; in the name of "chance," it was deprived of its verve. In any case, it was "on reprieve." It was not long for this world. It was condemned, irrevocably, to "thermal death."

Thanks to a conceptual revolution as important as that of relativity or quantum physics, scientists in our day are now able to appreciate the true value of the "moment." They are reconciled to the unpredictable, the novel, and, ultimately, to the idea of freedom.

To paraphrase Mallarmé and his swan imprisoned by the ice, one may say that the thaw has arrived and "today" has finally received its liberating "blow of drunken wing."[2]

The next three chapters will be devoted to the high points of this evolution. Several factors played an important role: the development of "chaos theory," the realization that the universe is expanding, progress in computer science, and the construction of superfast computers. Thanks to the contributions made by each of these elements, we can now appreciate, for its own special qualities, the flight of yellow butterflies over a field of colza.

DEMOCRITUS'S MESSAGE

"EVERYTHING HAPPENS ACCORDING to *chance* and *necessity*," wrote the Greek philosopher Democritus 2,500 years ago. This formula would enjoy an enormous success. It defined the two major poles around which different schools would rally all through the history of thought.

"All is necessity" implied that, thanks to a collection of dictatorial laws, the phenomena of nature and the events in our lives would be inescapably determined as what they are, and all intimations of freedom declared illusory. "All is chance" would imply, on the contrary, that nothing is determined. The fortuitous and the unpredictable would prevail completely.

Democritus, a shrewd thinker, was content to connect the two poles with a perfidious "and." He left to his readers the task of clarifying how "chance" and "necessity" *coexist* in the real world. It took over two thousand years to finally resolve this problem!

THE GOLDEN AGE OF THE "NECESSITY" POLE

IN THE HISTORY OF SCIENCE, the "necessity" pole had its hour of glory in the seventeenth and eighteenth centuries, thanks to the discovery of celestial mechanics. Associated with this glory are the names Newton and Laplace.

We can calculate the positions of the stars in the past as well as in the future. To the "celestial mechanics" who first performed these calculations, the solar system was a superb machine, a clockwork movement admirably calibrated for eternity.

In the Age of Enlightenment, the success of the theory of universal gravitation rendered the human mind giddy with its own powers. It seemed as though Newton's laws could serve as a model for all fields of inquiry. The whole of nature would be reduced to formulas. Cut down to size by mathematical equations, the universe would be open to human investigation.

A victim of this Newtonian conquest, "time" was put to death. The present moment was only an inconsequential coordinate in the uneventful unfolding of a "duration" reduced to the status of a mathematical parameter. The past and the future were hemmed in by a network of differential equations. For those who knew how to calculate, the mysteries of the world dissipated like the morning mist at sunrise. No one celebrated this development more impressively than Laplace. His text will be found in a note at the end of the book.[3]

If, regarding the airy flight of butterflies over a field of colza on a beautiful spring morning, we had asked the Marquis de Laplace the following questions — "Was this event destined to occur?" and "Could it have been predicted to the smallest detail?" — what would he have replied? If he was true to his world-view, he would likely have answered "yes" each time.

The pre-eminence of the "necessity" pole can be seductive to the human mind. But what it offers is a drab spectacle. Under its aegis, the adjectives "unexpected," "unforeseen," "novel," along with the nouns "inventiveness" and "creativity," would be reduced to sheer folklore. They would have to be stricken mercilessly from scientific vocabulary. The same would hold true for the words "freedom" and "fantasy." Chance would be unmasked. It would be seen only as an admission of ignorance, a lack of diligence in using to the fullest the human mind's enormous capabilities.

In their solemn splendor, the planetary orbits, tirelessly and endlessly retraced, present a perfect picture of monotony, boredom, and despair. They illustrate clearly the bankruptcy of "pure necessity."[4] They are difficult to reconcile with the inventiveness of nature, and they do not square well with the rich diversity of forms that flourished on earth through the ages.

THE GOLDEN AGE OF THE "CHANCE" POLE

THANKS TO DEVELOPMENTS in biology at the end of the nineteenth century, the "chance" pole also had its time in the sun. In the wake of evolutionary theory, randomness once again came into its own.

When we speak of "artificial selection," we mean that process by which a breeder of animals picks out the best of his livestock and controls their reproduction in such a way as to improve the quality. Darwin's genius was to imagine that a similar process took place in the natural world.

Unlike artificial selection, "natural selection" is not propelled by a superior authority's intelligent decisions. According to modern biology, it results from the blind behavior of a number of random genetic mutations. Certain mutations enhance the probability of reaching adulthood and having offspring. Individuals pass on this useful acquisition to their own children. And the line increases in number.

From this point of view, biological evolution is often seen as a huge lottery presided over by the god "chance." A brilliant biochemist of our time has defined "life" as "the result of a long series of accidents." For another, life "is nothing but a physico-chemical fluctuation that has endured for three and a half billion years."

The biologist Jacques Monod has helped popularize this approach in France. The title of his book, *Chance and*

Necessity, places Democritus's two poles on the same plane, but he does not treat them equally. "Chance" is favored over "necessity." Nothing, he writes, guaranteed that life would appear on earth, or that human beings would emerge from the animal world. These two phenomena, according to Monod, were extremely improbable. "Chance" alone decided.

THE CLUTTER OF RANDOM LANDSCAPES

THE COLD RIGIDITY OF PLANETARY ORBITS has served to illustrate the monotony of landscapes ruled by pure necessity. But what would nature look like entirely given over to chance?

Strictly speaking, of course, the question is meaningless. Physical laws cannot be broken. However, we do know of situations where their effect is minimal. The components of some landscapes are, for all practical purposes, a product of chance.

The distribution of meteoritic craters on the surface of the moon or Mercury provides a good example. The dimensions of the meteorites falling on the planetary surface, as well as the angles of their descent, obey no particular law. The distribution of the craters is perfectly random. It presents a cluttered picture.

The human mind can find nothing there to grab hold of, nothing in which it may take pleasure. Such is the character of a landscape shaped by chance alone.

The microscopic world of interstellar dust provides us with a second example. In the debris of exploded stars, specks of dust are formed from oxygen and silicone atoms ejected from the stellar furnace. Scattered in space, these specks are bombarded for millions of years by our galaxy's "cosmic rays."

Cosmic radiation is made up of particles — for the most part protons and electrons, but also, in smaller quantities, the nuclei of heavier atoms — that have accelerated almost to the speed of light. When one of these particles strikes a speck of dust, it penetrates deeply. It bores a tiny tunnel in the rock.[5]

In the laboratory, these specks can be "developed," like a photographic emulsion. On some, the bombardment has been particularly intense. The tunnels overlap and form an indescribable jumble. Like the lunar surface, this microscopic muddle illustrates the effect of random phenomena.

The rigidity of necessity's children and the muddle of chance's offspring both show us the extent to which these two poles, taken separately, are incapable of doing justice to the richness and splendor of the real world. To understand the appearance in the universe of structures increasingly organized and effective, we must seek further.

THE RECONCILIATION

TODAY, THANKS TO progress in scientific theory and numerical calculation, we have resolved the problem raised implicitly by Democritus's declaration. "Chance" and "necessity" are both crucial to the growth of cosmic complexity. We know how they are able to coexist peacefully and interact profitably. We can define their respective roles in this vast enterprise of organizing cosmic matter.

The authority of physical laws has not been called into question. The legislation remains, but its executive power is much less extensive than was thought in the eighteenth century. Our ability to know the future has been profoundly affected. The Marquis de Laplace could never have foreseen the curious look a nuthatch just shot me, through the open window of my office.

This partial randomness of events is nature's playground. It is there that it can create something new. It's in this natural habitat that we encounter freedom.

THE THEORIES OF CHAOS

LET'S TAKE A CLOSER look at the situation. Why is Laplace's dream utopian? What has changed since the Age of Enlightenment?

Very little, to tell the truth, but all of it significant. Laplace tells us that, to know the future, it's *enough* to know "all the forces by which nature is animated and *the respective situation of the beings that compose it.*" There's the rub. It's an impossible dream.

Over the last decades, many researchers have studied the problem. Their body of work has been labeled, not all that successfully, the "theory of deterministic chaos." To appreciate it properly, we must first make ourselves familiar with a number of new expressions, such as "sensitive dependence on initial conditions," "linear dependence," "nonlinear dependence," and "predictive horizon."

The situation may be illustrated with a simple example: the behavior of clocks. Not all clocks keep time with equal accuracy. Some are more reliable than others. Let's look at three different cases.

First of all, let's imagine an ideal clock that tells time perfectly. The first thing we have to do is set it. We telephone the "talking clock." It gives us the time to the second: "It is three hours, twenty-six minutes, thirty seconds, plus or minus a second."

In the case of a clock that is perfectly accurate, this uncertainty does not change. Twenty-four hours later, it tells the time with the same precision (plus or minus a second). After two, three, four days, the situation will remain unchanged.

In the real world, no clock tells the time with such fidelity. In their specifications, clockmakers indicate the accuracy of their product. Let's take the case of a *second* clock, that is accurate to *one second a day*. At the end of the first day, we will know the time within two seconds. (One of these seconds derives from the initial data, the other from the margin of error.)

If we don't reset the clock, our uncertainty will increase in "*linear*" fashion as time goes on. By "linear," we simply mean that it doubles each time the period under consideration becomes twice as long. After two, three, four days, our second clock will be accurate to within three, four, five seconds, etc. — plus or minus.

Let us suppose that the margin of error of our *third* clock, rather than remaining constant like the second (one second per day), doubles *each day* (one second after the first day, two seconds after the second, four seconds after three days, eight seconds after four days, etc.). Its progression is thus *nonlinear*.

How many days will it take before it is accurate to within only twelve hours? Answer: less than sixteen days. Sixteen days after our imaginary clock has been set, it will tell the time to within twelve hours. It will be perfectly useless.

This calculation introduces an important concept into our discussion: *the predictive horizon*. This is *the time period beyond which, in a given context, we can no longer foresee or affirm anything*. We may say that the "predictive horizon" of this third clock is sixteen days.

I have illustrated, in the drawing on page 90, the behavior of our three clocks. The degree of uncertainty is represented by the crosshatched "pie slices." The real time is to be found somewhere within these zones. For purposes of clarity, I have replaced "plus or minus one second" with "plus or minus one hour." In the drawings in the left-hand column, therefore, the real time is somewhere between eleven and one o'clock.

SENSITIVE DEPENDENCE ON INITIAL CONDITIONS

THESE CLOCKS WILL HELP US understand a concept that is central to our argument: "sensitive dependence on initial conditions." Here, the initial condition is the time as set, with an accuracy of plus or minus one second. Our three clocks do not have the same sensitivity to this initial data. Their predictive horizon is a reflection of this sensitivity.

The first clock does not increase the uncertainty established by the initial data, the second increases it in "linear" fashion, and the third amplifies it so dramatically that, after a short time, we know nothing at all.

The mode of amplification (nonlinear) of this last clock makes it very "sensitive" to the accuracy of the initial data. If we set the time to the nearest thousandth of a second, its predictive horizon increases to 26 days. To the millionth of a second, it reaches 36 days. It could last three months if we set the initial accuracy to 10^{-43} seconds (1/10,000,000,000,000,000,000,000,000,000,000,000,000,000,000).

This is the limit. Because of quantum effects, it's impossible to tell the time with any greater precision.[6] This limit, in return, gives us the absolute predictive horizon of our third clock: three months.

OF LAWS AND CLOCKS

LIKE CLOCKS, PHYSICAL LAWS have their own "predictive horizon."

These laws are expressed in *mathematical equations*. The equations tell us how the objects described by physics *change* through time. If we know the state of these objects (temperature, electric fields, etc.) at a given moment, the laws tell us what their state will be in the next moment. They instruct us as to the *difference* between the present instant and the one that follows. For this reason, they are called "*differential equations*."

These same words can be used to describe the behavior of clocks. They tell us, in conventional units (hours, minutes, seconds), how "the-time-it-is" changes with time. They indicate the *difference* between the present instant and the instant after. In this sense, clocks and differential equations work in the same way.

As long as we know the present well, physical equations will tell us the immediate future. This knowledge in turn can be used to evaluate what will happen in the next instant. Such was Laplace's strategy. For a long time it was thought that this strategy would enable us to predict *all* the future, from eclipses of the sun to the great migrations of wild geese over the arctic tundra as the days begin to lengthen.

The situation is identical to that with the clocks. Our knowledge of the present is necessarily imperfect. It is impossible for us to know with infinite accuracy where the moon is at a given moment. Our telescopes give us its position with a certain margin of error. We will say, for example, the moon is in a certain direction *plus or minus* a millionth of a second of an arc. Once again, we have the problem of sensitivity to initial conditions.

Let's consider, again as an example, the movement of the Earth around the sun. Set aside, for the moment, the presence of other bodies in the solar system. In the seventeenth century, Newton devised the equation that describes this movement. This equation is linear (we'll see a little later the exact meaning of this term). Like our second clock, this equation depends very little on the accuracy of the initial data. It could trace indefinitely the position of the Earth.

But there is more to the solar system than the sun and the Earth. There are other planets that influence the Earth's progress through space. We must, in principle, take them into consideration and account for them in our equations. In reality, these planets are much less massive than the sun. On the time scales that interest us, their influence is weak. In practical terms, we ignore them.

The situation would be different if Mercury and Venus, for instance, were much larger. Wc would include them in our equations, which, as a result, would become very complicated. Their sensitivity to the initial conditions would parallel that of our third clock. The predictive horizon — beyond which any prediction of eclipses would become impossible — would be very brief. We would talk then of chaotic orbits.[7]

Such cases do exist in the solar system. Data from the Voyager space probe have shown that certain satellites of Saturn, revolving close to the rings, have particularly strange movements, impossible to predict.

From this analysis we have learned that only linear equations (of *low* sensitivity to initial data) allow us to predict the future in the long term. Generally, sensitivity to initial conditions increases rapidly as more bodies interact.[8] And in almost all concrete situations, this number is sufficiently high to render the equations extremely sensitive to initial conditions.

The luck (or the misfortune) of the celestial mechanics is to have stumbled on the *very rare* exceptions to this rule. This accounts for their premature optimism where the predictive capacities of the sciences are concerned. In the real world, the number of "future states" that lend themselves to long-term evaluation is infinitely small. Having come to terms with this fact, scientists today are more realistic and more modest . . .

RETROACTIVE LOOPS

"LINEAR DEPENDENCES," "nonlinears," "sensitive dependence on initial conditions," "predictive horizons" — to better understand our problem, we must acquaint ourselves further with these concepts. Some real-life examples will show us how they interact.

The heat given off by a radiator increases the temperature of a room. A certain quantity of heat (the cause) induces a certain increase in temperature (the effect). Under some conditions, if I double the heat given off, I double the temperature (in absolute degrees). We then say that the temperature depends *in linear fashion* on the heat produced.

This description does not apply, however, to a forest fire. The heat of a single match can set ablaze a whole region and result in extremely high temperatures. The reason is well known: the (weak) increase in temperature caused by the match flame is enough to set fire to a few dead branches. These new fires give off even more heat, raising the temperature again, etc.

It's clear that in this case the intensity of the effect influences, in its turn, the intensity of the cause. This is called a "retroactive loop"; the cause produces the effect and the effect acts on the cause, which acts on the effect, etc. In a forest fire, the temperature depends *in nonlinear fashion* on the heat given off by the match.

If the effect does not influence the cause (linear dependence), the equations are not very sensitive to initial conditions. That makes it possible to calculate the temperature's evolution over a long period.

If, on the contrary, the increase in temperature results in the production of even more heat (nonlinear dependence, retroactive loop), then, by amplifying the cause, the effect creates a strong sensitivity to initial conditions. The predictive horizon for the temperature's evolution will be very short. The behavior of a forest fire is difficult to predict.

Now, as it happens, problems in physics abound in nonlinear equations. The real world is full of "retroactive loops" . . .

THE "BUTTERFLY" EFFECT

TO CALCULATE THE PLANETARY ORBITS, we need only consider the influence of a small number of celestial bodies. To study meteorology, we must account for the presence of billions and billions of air molecules. No one, of course, wants to follow the trajectory of every molecule in the atmosphere. We're interested in "mean values," such as temperature, wind, and rain patterns in a particular region. Even these — as we all know — can be predicted with only relative accuracy by professional meteorologists . . .

If we were to increase the power of our computers, would that be enough to improve the quality of our predictions? For a long time we thought so. But this was to ignore the *extreme sensitivity* of aerodynamic equations to initial conditions.

Let's try, for example, to calculate the weather one year from now on the same date. First, we would have to enter into the program *everything* we know about the state of the atmosphere at the present moment: temperature distribution, clouds, winds, etc. We then set the machine in motion and calculate. Suppose the result indicates fair weather, no clouds.

Now it so happens that somewhere on the planet a butterfly has taken flight just as we began to calculate. We didn't account for the slight breath of air created by the movement of its wings. We have to start all over again and include this new piece of initial data. Surprise! The atmospheric effects of this flight have a profound influence on our calculations. We have to revise our predictions for the following year: it's going to rain!

Must we then give up on weather forecasting? No, not if we can be happy with the short term. For tomorrow, it works quite well. For three days hence, it's already a little less certain. After two weeks, you're just as likely to be right if you toss a coin . . .

We can, in this way, for each scientific field, posit a "predictive horizon" beyond which the future is unknowable. We can predict eclipses thousands, even hundreds of thousands of years into the future, but there is always a time beyond which we cannot go. The predictive horizon for meteorology is about a week. It will do no good to buy a better computer. The butterflies will render it useless.[9]

A SUBTLE DOSAGE

RECENT DEVELOPMENTS IN PHYSICS, and in particular, the birth of "chaos theory," have taught us how laws can coexist with the dimension of play in nature. And how the simultaneous presence of the two poles, "chance" and "necessity," is crucial to inventiveness and creativity in the world.

The role of laws is not to decree an inexorable future. Their executive power in the long term is clearly not up to that task. With present conditions as a starting point, they're content to establish an assortment of possibilities, all compatible with current legislation. Beyond the predictive horizon is freedom's domain.

When one of these possibilities becomes reality, it in turn opens onto a new assortment. These assortments branch out, graft onto each other, and multiply into an infinite number of possibilities, almost impossible to predict.

It's such a sequence of events — involving both law and chance — that is responsible for the freshness and verve of our todays. Through their subtle interaction, the two poles collaborate in the unfolding of the world.

And so, seen in its true complexity, the universe can be likened neither to the tidy arena of planetary orbits so dear to the "wise men" of the Age of Enlightenment nor to the sterile clutter of planetary surfaces riddled with meteoritic scars.[10] Rather, we see before us the palette of an imaginative painter, forever at work producing novel effects.

6

Thermodynamics and the Flight of Butterflies

IN CHAPTER FIVE WE SAW HOW the development of chaos theory gave us insight into the words of Democritus. Two phenomena that seem to be mutually exclusive, chance and necessity, not only coexist but together make for a varied universe and the liveliness of the present moment.

In Chapter Six, we'll see how the discovery of an expanding universe, in turn, gives us a new slant on our frolicking butterflies.

Nineteenth-century physicists regarded the future of the universe with a jaundiced eye. At that time, a new concept, "entropy," had made its way into scientific language.

I will deal with only one aspect of this complex and difficult notion: within a given medium, entropy is a measure of disorder, or, more exactly, the absence of organization. Applied to the universe, thermodynamics affirms that *the entropy of all matter in the cosmos is continually increasing*.

This notion and this law have given rise to hasty extrapolations that long served to justify a pessimistic view of the world. Sooner or later, our "todays" would be hopelessly doomed. The inexorable deterioration of the cosmos implied that, at some point, *nothing more would happen*. From

Wittgenstein's vantage point, the universe "would no longer exist."

The discovery that the universe is expanding called into question these somber predictions. When, in particular, we factor in the effect of gravity on matter in the universe as it cools, we get a clearer understanding of the cosmic conditions that guarantee the universe a long life. I'll begin this part of the book by contesting three popular assumptions that claim to draw support from thermodynamics.

LIFE IS NOT EXEMPT FROM THE "RULE OF ENTROPY"

IT'S OBVIOUS THAT LIVING MATTER is more organized (less entropic) than inert matter, out of which it emerged when our planet was young. From this fact, many people quickly assumed that life was incompatible with the laws of thermodynamics and that living beings had to be exceptions to the rule of increasing entropy.

This interpretation would support the idea that life is an extremely improbable phenomenon, almost a "miracle." Out of this improbability there developed an argument against the hypothesis of extraterrestrial civilizations. "Man knows at last that he is alone in the uncaring immensity of the universe, out of which he emerged by chance," wrote Jacques Monod.

What should we make of this? Life is perfectly compatible with thermodynamics. The second law of thermodynamics allows for the creation of order and organization *in a given sector of the cosmos* as long as it is not closed in upon itself. It must be in communication with the outside world. This increase in order must also be outweighed *by a greater growth of disorder in the universe as a whole.*

Order is not created only in the realm of the living. It exists wherever matter is being structured. When two atoms of hydrogen link up with an atom of oxygen to form a molecule of water, we have organization (the molecule is more "organized" than free atoms). This chemical reaction is accompanied by the emission of luminous photons. These photons are projected into intergalactic space. They represent, for the universe, increased "disorder."[1] The second law comes into its own with this emission. In a static universe (not expanding) these photons would gradually heat up the heavens, posing a threat to the future of life. This threat is neutralized by expansion. We'll pick up on that point later on.

Similar phenomena, although much more complex, occur when life appears, and at sunset swallows flit over the pond at Malicorne. All these living organisms give off heat. This thermal energy, in the form of infrared radiation, is absorbed into immense interstellar space, thereby paying its dues to the rule of entropy. Life is not exempt from the laws of thermodynamics. Its existence is neither more nor less "miraculous" than the formation of water from hydrogen and oxygen atoms.

LIFE IS NOT DESTINED TO DISAPPEAR

OUR PRESENT UNIVERSE is far from being isothermal (at the same temperature everywhere). Between the cores of stars — at millions, if not billions of degrees — and interstellar space — at less than three degrees absolute — we find a whole range of temperatures.

When thermodynamics was a new idea, it was assumed, prematurely, that these differences in temperature would inevitably diminish and disappear. Sooner or later, cosmic matter everywhere would reach the same temperature, rendering useless all the energy in the universe. The universe itself would be condemned, without appeal, to "thermal death."

Modern astronomy presents us with a completely inverted picture of the universe. From an almost completely isothermal state, fifteen billion years ago, matter has progressed to its current condition, in which there are significant variations in temperature. Where did the pioneers of thermodynamics go wrong?

It took some time before gravity's effect on thermodynamics in the universe was properly understood. It is in the stars that this role is played out.[2] In the primordial soup there are no stars. They appear when gravity, at work within cosmic nebulae, causes them to contract and heat up. Thanks to these innumerable contractions, the universe gradually emerged from its initial isothermal state. Today, billions of degrees separate the temperatures of stellar crucibles from those of cold stars adrift in space.

Here again, nothing exceptional, nothing miraculous. It's the law of entropy itself, applied in the context of an expanding universe, that is responsible for the appearance of these differences in temperature so important to the origins of life and the liveliness of our todays.

WE CANNOT AFFIRM THAT THE UNIVERSE IS MOVING INEXORABLY TOWARD A STATE OF MAXIMAL DISORDER

A THIRD DUBIOUS claim goes like this: even if the universe is not moving toward thermal death, it is, in any case, doomed by the law of increase in global disorder.

This idea derives from the superimposition of two assertions: (a) all "events" in the universe, such as the stately flight of herons over marshland, can only "happen" *if* they contribute to the increase of entropy in the universe; (b) the day will come when this entropy will have reached its maximum value and will be incapable of any further growth. Conclusion: at this point, there will be no more "events."

The first assertion is correct. The second would be true if (as was once believed) the universe were static. It is no longer true for an expanding universe.

Let us compare the universe to a basin into which we're pouring water.

An "ordinary" basin would eventually become full. The idea of a universe that could one day reach its level of *maximum entropy* would be comparable to an ordinary basin. This would be the situation if the universe were static (cf. p. 104).

But we can also imagine a different basin, with sliding walls, that could constantly increase its own volume. Such a basin, far from overflowing, would always be able to hold the water poured into it.

An extensive study of the properties of gravitational force shows that the universe behaves not like an ordinary basin but like our second basin. *It can increase indefinitely its entropy.* A detailed demonstration of this point may be found in Appendix B.

This limitless ability to increase its entropy is what guarantees life on earth. Thanks to it, the universe can at each moment preserve the freshness of its "todays" and the exuberance of a flight of butterflies over fields of colza.

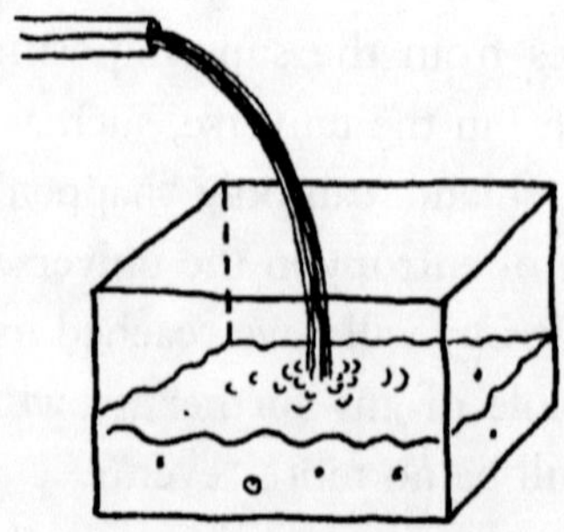

Just as an ordinary basin into which water is poured will eventually become full, so a static universe will in time reach a state of maximum entropy.

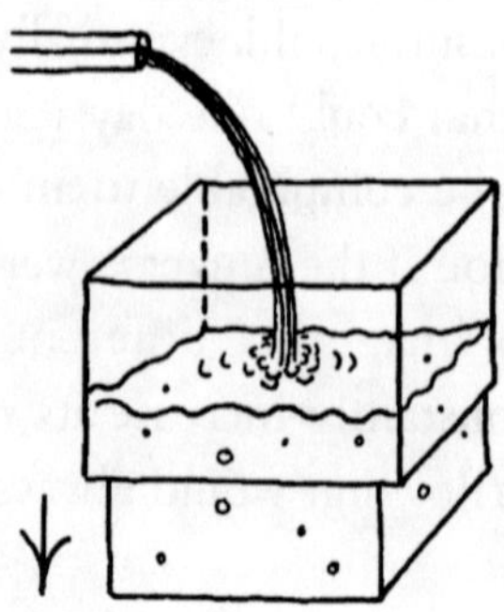

Just as a basin with a sliding bottom will never fill, so an expanding universe will never reach a point of maximum entropy.

BUTTERFLIES AND THE EXPANSION OF THE UNIVERSE

WE ARE NOW GOING TO STUDY in a more detailed fashion the close ties linking the expansion of the universe to an increase in complexity. This subject was treated at length in *The Hour of Our Delight*. But before we investigate further the freshness of the present moment, we have to briefly retrace our steps.

The ancient Greeks were the first, as far as we know, to study nature through observation and scientific theory. Let's imagine that Aristotle, one of the pioneers in this enterprise, were to return among us and ask: "What do you know about nature that we didn't know in our time? What have you learned that is new since we walked upon the earth?"

It would be simple, but unfair, to show him the millions of books stored on the shelves of our scientific libraries. Let us rather sum up, in a few words, our current thinking on the structure of matter.

We could answer Aristotle's question with two sentences: 1. "Nature is structured like a language"; and 2. "A pyramid of complexity has arisen over the ages."

NATURE IS STRUCTURED LIKE A LANGUAGE

A WRITTEN LANGUAGE is made up of letters, words, sentences, paragraphs, chapters, books, collections. The basic formula is combinative. Words are combinations of letters, and a combination of words gives us sentences.[3]

Here we encounter an important concept: the "emergent property." The combination of letters in a specific order results in the appearance of something "new," something we can't find when we look at each element separately. The word "black" evokes an image totally foreign to its individual letters. This image "emerges" not from the letters themselves but from their association in a given order. The meaning of every word in the dictionary is an "emergent property" resulting from this combinative process, according to rules whose origins are both historical and conventional.

These emergent properties come into play at every level. The meaning of the sentence "The night is black" arises from a juxtaposition of words in conformity with rules laid down by syntax and grammar. And the juxtaposition of sentences transmits a still wider meaning, a message of even greater scope.

We can therefore use the image of a pyramid with superimposed "alphabets" (letters, words, sentences, paragraphs, etc.) to describe a written language. Each element at a particular level is composed of elements from the level below and makes up the elements of the level above. Words are the "letters" of sentences, sentences are the "letters" of paragraphs, and so on. The basic principle of this construction is, once again, a combinative process that generates emergent properties.

Our first statement, "Nature is structured like a language," tells us that nature functions in the same way. (For the sake of simplicity, I'm talking here only of a written language.) We've come to realize that this formula (writing), invented by human beings some five or six thousand years ago, has been used in nature for fifteen billion years.

Chemists deserve the credit for having first discovered this process in nature, on the level of atoms and molecules. Thanks to Lavoisier, Priestley, Dalton, Gay-Lussac, and many others, it was learned, during the eighteenth and nineteenth centuries, that molecules are like words whose letters are atoms.

The most important task of the chemists was to establish the list of chemical elements that formed the totality of natural substances — in other words, to identify the *alphabet of atoms*. This list is made up of a little over a hundred atoms. Traditionally, these elements are arranged in a "periodic table," found on laboratory walls and in the classrooms of our educational institutions.

To bind together the letters of a word, nature draws on a number of forces. Atoms are "welded" into molecules by *electromagnetic force*. As its name indicates, this force is responsible for all electric and magnetic phenomena. It controls all chemical reactions and all the effects of light.

At the beginning of the twentieth century, physicists adopted an approach similar to that of the chemists. If atoms are the letters of molecules, they are also the words of a new alphabet, that of nucleons. There are many sorts of nucleons. Two of them, the proton and the neutron, are components of nature's stable nuclei.

Carbon, for example (more precisely the nucleus of the carbon atom, surrounded by an entourage of electrons), is made up of 6 protons and 6 neutrons. Iron is composed of 26 protons and 30 neutrons. Uranium has 92 protons and a little over 140 neutrons, depending on the type. We are now familiar with over a thousand different nuclei with protons and neutrons combined in specific proportions. Here, the binding agent for the "letters" is the nuclear force, which holds together protons and neutrons within the atomic nucleus.

Nucleons themselves are words in relation to particles called *quarks*. There are six sorts of quarks, but only two are stable, the *u* quark and the *d* quark. A proton is, for all practical purposes, composed of two *u* quarks and one *d* quark, while a neutron is made up of one *u* quark and two *d* quarks. Once again, the nuclear force is at work here. It binds the quarks into nucleons just as it binds the nucleons into a nucleus.

The twenty-six letters of the English alphabet are by definition elementary. They do not form "words" for another alphabet. What about quarks and electrons? We think, without being certain, that these are elementary particles. Accelerators more powerful than those at our disposal today may, in theory, shed light on that question.

By clarifying the fundamental role of cells in living organisms biologists have made their contribution to our pyramid of complexity. An alphabet of about two hundred cell varieties is sufficient to spell out our huge assortment of plants and animals. Biochemistry, in turn, has revealed that living cells incorporate an alphabet of giant molecules (proteins, DNA, etc.). The electromagnetic force holds together and ensures the functioning of cells and organisms.

In parallel with the pyramid of written language (letters, words, sentences, paragraphs, chapters, etc.) we can today raise up the pyramid of nature's superimposed alphabets.

The lowest level is home to the nuclear force, responsible for combining quarks into nucleons and nucleons into nuclei. Higher up, we enter the territory of the electromagnetic force, in charge of the formation and workings of molecules, cells, and living organisms.

Why a pyramid? The organization of matter affects *only a fraction of the matter in the cosmos*. This fraction shrinks the higher we climb on the pyramid. The total mass of all living beings, for example, is but a tiny part of the mass of the earth. Almost all of the universe occupies the lower levels of the pyramid. Here we can see reflected the "rule of entropy," described in the previous section. The local organization of matter is necessarily compensated for by a more significant overall disorganization. The universe cannot transform itself entirely into human brains . . .

Nature is structured like a language.

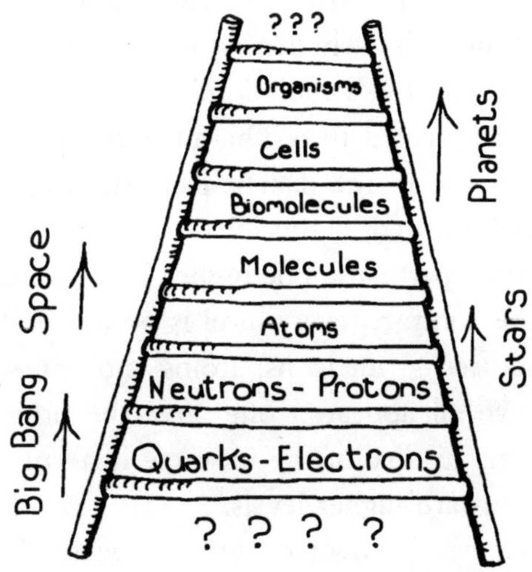

The pyramid is constructed in time.

We have now explained the meaning of our first key statement: "Nature is structured like a language."

THE CONSTRUCTION OF THE PYRAMID

THE SECOND KEY STATEMENT in our message to scientific pioneers comes to us from astronomical and cosmological research. It goes like this: "The pyramid of complexity was erected in the course of time."

The discovery of the galaxies' directional movement is what put us on the trail. From this there emerged a picture of a universe that was not static, as had long been believed, but evolving. The galaxies are moving away from each other. In the past, they were closer together.

Starting with our observations of the galaxies' movements, and with the help of our knowledge of physics, we can travel back in time. We can reconstruct early conditions in the cosmos. It appears progressively denser, hotter, and brighter.

The detection of "fossil radiation," emitted fifteen billion years ago, has confirmed this. This radiation gives us a picture of the universe in the distant past, showing that all of matter was concentrated at the lowest level of our pyramid, that of elementary particles. The higher levels were empty.

Today, on the contrary, every level is occupied. The corresponding structures — nucleons, atoms, molecules, cells, organisms — have all appeared one after the other. We can thus describe the history of the universe in terms of the ascent of matter toward higher levels.

Each science has its own chapter to write. Physics describes for us the formation of nucleons out of the original soup. It tells us about the birth of stars, under the aegis of gravitational force, and the fusion of atomic nuclei in the cores of fiery stars. Chemistry tries to explain how molecules were formed in interstellar space, in the atmosphere, and in the earth's primitive ocean. Biology makes its contribution at the higher levels of the pyramid, those occupied by cells and multicellular organisms. Thus, over the ages, in the heavens and on earth, natural forces have raised up the pyramid of cosmic organization.

EQUILIBRIUMS ARE STERILE

WE ARE STILL MISSING one essential ingredient that would give us an adequate picture of how matter is organized. The role of natural forces, as we have seen, is to "bind" cosmic structures together and produce a variety of forms. However, the end product of their activity could have been very different: a sterile world, dull, and with no variety. What were the conditions that led to *diversity* rather than *monotony*?

The "dream" of the nuclear force was to transform all matter into iron. Every substance — hydrogen, uranium, or strawberry jam — on which this force might exercise to the limit its formidable cohesive powers would inevitably turn into iron.

In the same way, if we were to give "carte blanche" to the electromagnetic force, proteins, DNA, and life would be absent from our world. We would find only simple molecules (water, carbon dioxide, methane, ammonia) and noble atoms (helium, neon, argon, etc.).

Were the force of gravity not checked, stars and galaxies

would contract into "black holes," dead stars curled up on themselves.[4]

The universe would be a very different place had the natural forces been able to exploit to the limit their capacities for binding matter together. No "today," no events, no steel-blue dragonflies in the green countryside.

The most ancient picture of the universe is one of monotony, a uniform soup of elementary particles. Later, the nuclear force went into action and produced a variety of atomic nuclei. Later still, the electromagnetic force got to work and "grafted" onto these atoms a much richer variety of natural molecules. Today, after fifteen billion years, it's easy to see that these forces did not have "carte blanche." They did not exhaust their possibilities. The threats they posed to the richness of the universe came to nothing. Why?

Because the universe was expanding! A static and unchanging world would *necessarily* be monotonous, because the "dreams" of these forces would have time to materialize a hundred, a thousand, a million times over!

FORMS OF CRYSTALLIZATION

A FEW SIMPLE EXPERIMENTS will help us understand this fundamental point.

In a jar, we cool some hot water oversaturated with salt. *If the cooling is rapid*, we will find on the bottom of the jar a tangle of thin, delicate needles. A beam of polarized light, projected laterally through the needles, will produce a multi-hued rainbow.[5]

Let's repeat the experiment. The needles will reappear but

in a different configuration. There is no limit to the variety of possible entanglements. The same patterns never reappear.

A new series of experiments. This time let's put the salt water in a double-boiler and cool it *as slowly as possible*. The result: a block of salt fitting closely the form of the jar. No polarized light can make it iridescent. Try it again a hundred times, a thousand times: the same block, perfectly predictable, perfectly regular, will reappear before our eyes. No procession of different forms, as observed in the first series of experiments!

Why does the speed of cooling have such an influence on the result? If it's slow (the second series of experiments), the atoms *have all the time they need* to occupy the available space on the crystals as they form. The irregularities disappear and the gaps are filled. The surfaces become smooth and necessarily produce the predictable block of salt. We say then that the reactions are in *equilibrium*: everything that *can* happen *has the time* to happen.

On the other hand, when the cooling is rapid, there's a stampede. The atoms grab on where they can. In their haste they produce outcroppings that are never smoothed down, cavities that are never filled in.

Now let us compare crystal formation to the appearance of the first atomic nuclei in the cooling universe. At the very high initial temperatures, there were only elementary particles. If the universe had cooled slowly, cosmic matter would have remained indefinitely in a state of equilibrium. The natural forces would have exploited their capabilities to the limit. They would have had the time to bind everything together. Endless monotony would have reigned in the universe.

The temperature of an automobile cylinder lowers when the piston pulls out. In the same way, the universe cools as it expands. Observation and theory both show that this expansion was *too rapid* to maintain a state of equilibrium. The forces didn't have the necessary leisure to achieve everything in their power. Iron exists, but it's rare: today, there is only one atom of iron for every hundred thousand atoms of hydrogen. With stellar activity, this number is still growing, but more and more slowly. All indications suggest that we will never have a cosmos of pure iron.[6]

The universe was saved from sterile equilibrium by the speed of its cooling. The disequilibrium thus produced left the cosmos with the opportunity to become ever more rich in splendor and diversity. It is the protector of the unforeseeable, the novel, and of our todays. It allows migrating birds to take off in large flocks when the first harbingers of autumn come round again.

7

Butterflies and Time's Arrow

IN THE LAST TWO CHAPTERS, we described science's new approach to the real world. We saw how the development of "deterministic chaos theories" reconciled two seemingly incompatible ideas, "chance" and "necessity." We now have a better understanding of how these two elements work together and complement each other to stimulate creativity in nature.

The discovery of the expanding universe has also changed our view of the future and of the fate of cosmic complexity. Thermodynamics, revised and corrected to take account of this expansion, gives us a very different view of time's role in the history of the universe. "Time" is very much a central concern as we speculate on our todays — their pristine present and their future. This seventh chapter will be entirely devoted to the subject of time.

"What, then, is time? I know well enough what it is," said Saint Augustine, "provided that nobody asks me; but if I ask what it is and try to explain, I am baffled."

We tend to distinguish between different sorts of "time." First, there's "intuitive time," which we feel in our bodies and which, like Saint Augustine, we have such difficulty defining. There's also "physicist's time," which is entered into equations to reflect the behavior of matter when it conforms to nature's laws. Finally, there's "cosmological time," that of the Big Bang and cosmic evolution. We'll see how difficult it is to reconcile these concepts.

One basic difference concerns time's "reversibility." This subtle notion will be illustrated a little later with appropriate examples.

In his *Politics*, Plato tells us that in the age of Chronos, father of Jupiter, "time in the world was reversible, moving sometimes in one direction, sometimes in the other. What was the beginning at one moment could become the end in another, and vice versa."

Like that of Chronos, the physicist's time is "reversible." In physical equations, there is no distinction between present and future.

However, there is a basic difference between time in physics and intuitive time. For us, space and time are in no way equivalent. In space, we are free. We can go forwards, backwards, to the left or right, up or down. In time we do not have this freedom. We can't choose "to go to yesterday or tomorrow." Time carries us along ineluctably, like a moving train. We can't get off. We move inexorably from the past (from our memories, where we can no longer act) toward the future (open to all possibilities).

Our current understanding of the history and evolution of the cosmos does not present a picture of "cosmological time" that is reversible. Like our bodies, the universe seems to be evolving in one unchanging direction. How do we reconcile the reversible time of physics with an irreversible intuitive and cosmological time?

This discussion is at the center of a lively debate among physicists and astrophysicists. It revolves around a subtle notion: time's arrow.

I'll try to describe the situation using a little story. Then some preliminary analysis will help us understand why this arrow, so rooted in our intuition, has been declared illusory by physicists. We'll need a second look to detect the weaknesses in the argument on which this claim is based. We'll then see how the presence of predictive horizons, in an expanding universe, enables us to reconcile the three "times."

IDENTIFYING THE ARROW'S DIRECTION

AN EXTRATERRESTRIAL (a real one!) arrives from a world where there is no time. He wants to understand what the word means. To give him some concrete experience, we're going to teach him how to correctly identify time's arrow.

The game consists in showing him filmed sequences, which we might run either from the beginning or the end. We want him to be able to tell whether, in the projected sequence, time is flowing normally or in reverse.

First sequence: a billiard table seen from above. A ball enters at the left, crosses the green surface, and exits at the right.

The same sequence is then rerun in the opposite direction. The ball enters at the right and leaves by the left. Which is the "true" sequence and which is reversed?

A ball crosses the billiard table from left to right.

Reverse sequence: the ball crosses from right to left.

Our visitor can learn nothing from this sequence. There is no clue as to which is the true sequence and which is running backwards. It's as though time's arrow has left no trace. The movement is *reversible* in terms of time; for the ball, past and future are interchangeable. As in Plato's mythic world, the beginning can be the end and vice versa. If we know the past, we know the future.

Let's move on to another sequence. The setting is the same. Two balls converge, make contact, and move off in different directions.

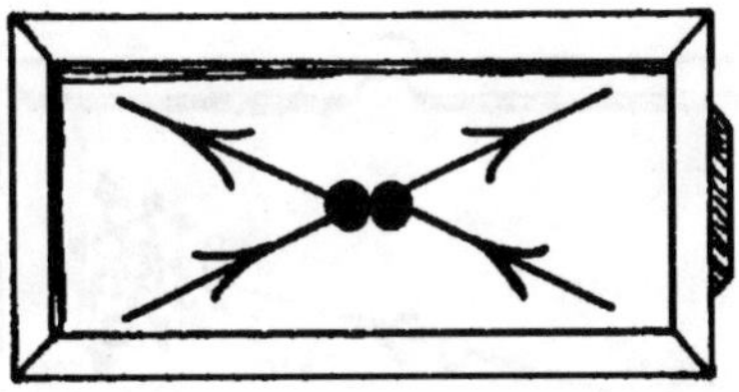

Second sequence: two balls collide.

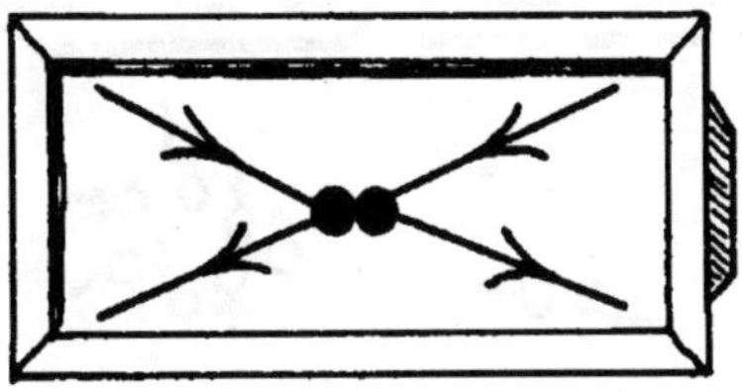

Reverse sequence.

When the sequence is run backwards, the directions are inverted, but the scenario has not changed. Here again the movements are reversible. Time's arrow is not operative.

In a third sequence we have the billiard table on which ten multicolored balls are arranged in a triangle.

A white ball arrives, strikes the triangle, and disperses the billiard balls in a number of directions. The reverse sequence shows ten colored balls converging on one corner of the table. They come to rest in a triangle. A white ball escapes and rolls into the opposite corner.

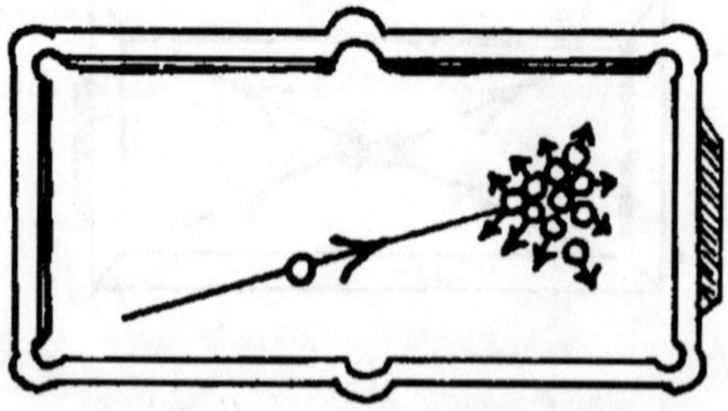

Third sequence: balls are dispersed.

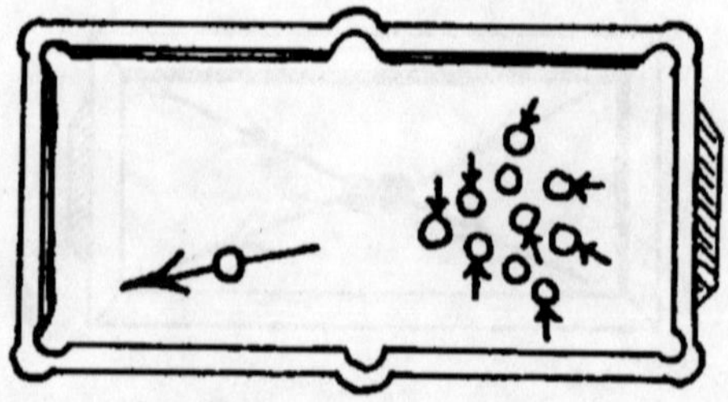

Reverse sequence.

Now it's easy to get our bearings. The first sequence is more natural, more believable. The second, of course, is not impossible. Such coordinated movements could take place. But they are extremely *improbable*.

Let's put it differently. There are a thousand ways for the colored balls to scatter in space (multiform movement). There is only one way for the white ball to make its way from the corner toward the triangle (simple movement). It's more probable that one would proceed from a simple movement to a multiform movement. The opposite could happen, but so rarely that we could bet against it without taking much of a risk. Multiform movements generally give rise to other multiform movements. It's with the words "probable" and "improbable," therefore, that we'll try to enlighten our visitor.

A fourth sequence will complete the lesson. On the screen, we show a drop of ink as it falls into a glass of water. It dilutes and spreads evenly. The water becomes a uniform gray. Our student, if he has understood what we have taught him, will have no difficulty distinguishing between the real sequence and the sequence in reverse. Here we're dealing not with eleven objects on a billiard table but with about 10,000,000,000,000,000,000,000 molecules. The probability of seeing the drop form on its own is so small that we need not take it into account. The sequence "multiform movement becoming a simple movement" has almost no chance of happening.

NOTHING MORE?

"COMPLEX" EVENTS (the dilution of a drop of ink) are "arrowed," while "simple" events (a collision of two bodies) are not. Such seem to be the conclusions that emerge from this discussion.

These conclusions could enable us to reconcile physical time with intuitive time. The reasoning would go as follows. Each *complex event* in reality could be broken down into a large number of *simple events*. A billiard game is a series of collisions between balls. The dilution of a drop of ink may be seen as an uncoordinated series of movements on the part of ink particles in a container of water. Cooking an egg comes down to a host of discrete chemical reactions in a frying pan.

Isolated and taken separately, these simple events resemble our first two sequences. They do not seem to have been affected by time's arrow. How is it that this arrow appears when we observe the event *in its entirety*, that is, when we look at these simple events as a larger series?

The sequences previously screened suggest an answer. *Time's arrow operates only in a transition from the less probable to the more probable.* In fact, complex events *are just as reversible* as simple events; they are simply less probable. If we are able to identify the right sequence, it is simply because we have learned from experience to distinguish probable from improbable events. Nothing more. If that's the case, then the irreversibility of intuitive time is pure illusion.

ANOTHER SHADOW CAST OVER TODAY

IT HAS TO BE UNDERSTOOD that this seemingly innocuous "nothing more" poses another threat to today's "freshness." For if we admit that complex events are reversible — even if the probability of such a reversal is small — then we must also admit that the passage of time creates nothing new.[1] In fact, the word reversible implies that what is in the future could just as easily have been in the past.

From this point of view, our ignorance of the future bears only on the question of *which* state is going to follow on the present one. Even if we can't know what is *going to happen*, in theory we should be able to know everything that *could happen.* (For example, when I toss a coin I know in advance that the result will be *heads* or *tails*, even if I don't know which to choose.) The range of possible states would be set and knowable. Our sense of the present as something pristine and fresh would be an illusion. *Creativity* would play no role in the world.

S.O.S. COMPUTERS

LET'S TAKE A CLOSER LOOK at the situation. The reasoning in the last two paragraphs is based on two assertions: 1. simple events are reversible; and 2. complex events — made up of simple events — are, despite appearances, also reversible.

Is this really true? In the following pages, we'll see that both assertions are questionable. *In the real world, neither complex nor simple events are exactly reversible.* Our first argument takes us back to "predictive horizons" and chaos theory, the second to the expansion of the universe. These two key elements will show us that these assertions are not exact, only approximate. It's their approximate nature that opens the way to a reconciliation.

Here, we look to our powerful computers for help. Who would have thought that these faithful servants, fast but stupid, would prove useful one day in a philosophical discussion? That they would shed light on the problem of freedom?

Let's reexamine the sequence with the drop of ink. First, we'll write out the equations for the movement of ink particles among molecules of water. The computer will enable us to reconstruct the progressive dilution of the ink. We project the result on a screen.

As in real life, we see the colored particles, at first confined in a small drop, then diffused throughout the liquid. The screen becomes gray.

For the opposite sequence, we now reverse the movement of each of the particles: "Reverse all engines."

We might think that, in retracing in the opposite direction their earlier path, thc ink particles would end up re-creating the initial drop on the screen. Nothing could be further from the truth.[2] *The drop does not reappear!* The screen remains gray.

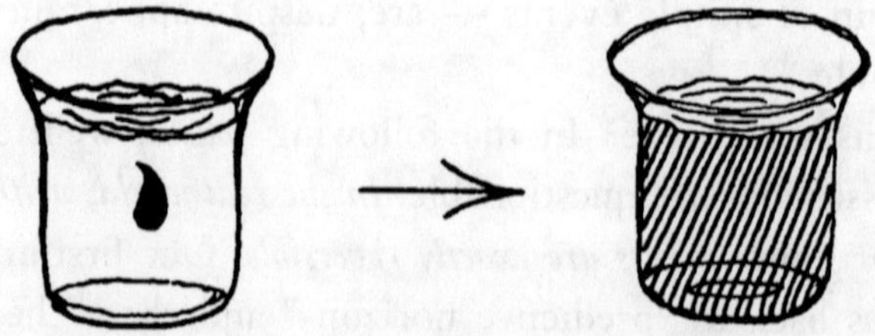

The drop of ink is diluted: the solution becomes uniform.

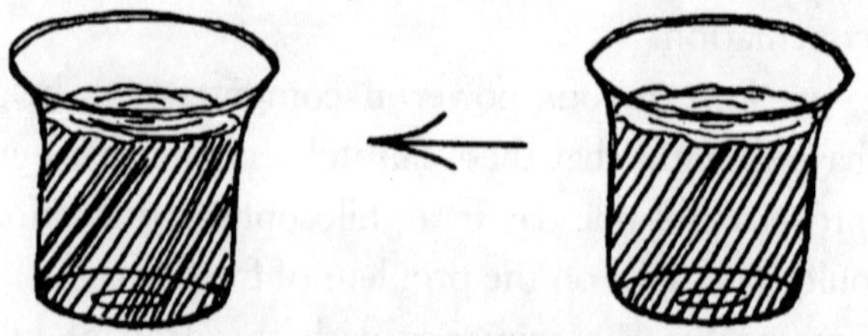

The drop of ink does not reform.

Contrary to what we might have supposed, the sequence of the drop of ink is not reversible. What has happened?

Here we encounter once more the problem of "sensitivity to initial conditions." Our computer *could* correctly plot the path in reverse — and reconstitute the drop of ink — as long as we could provide it with the position and exact speed of each of the particles. Now, because of the tremendous number of bodies interacting, the equations for the ink particles are *very sensitive* to initial conditions. Their predictive horizon is very short. It falls far short of the time required to reform the drop of ink. The screen remains obstinately gray.[3]

The computer provides us with proof: beyond the predictive horizon, complex events in nature are not reversible. However "deterministic" they may be, the equations that describe their movements cannot foresee the future in the long term. Their future is only *partially* inscribed in their present.

We may sum up the situation in the following way: *the equations are deterministic, but the solutions to the equations only approximately so. And beyond the predictive horizon, not at all.*[4]

Our focus suddenly shifts to the simple events. Are they really reversible? The answer is still negative. The difficulty this time is not with the predictive horizons but with the expansion of the universe, which enables us to link physical time with cosmological time.

Let's take a closer look at the sequence of the two balls that converge, collide, and separate.

To adequately describe the event, we must take into consideration the fact that the balls are not simple mechanical objects. They're composed of atoms that interact with the rest of the universe. We have to account for all these interactions; it's there that time's arrow comes into play.

Like all objects, these balls emit radiation proportional to their temperature. This radiation, made up of infrared photons, leaves the balls and disperses in space.

Moreover, we have to take into consideration the existence of fossil radiation composed of millimeter-wavelength photons (whose energy is much weaker than that of infrared photons).[5] All along their path the balls absorb these photons. As it expands, the temperature of the universe diminishes (very) slightly. The millimeter-range photons absorbed are a little less numerous and have a little less energy at the end than at the beginning.

The reversed sequence is different from the real sequence. We see the infrared photons arriving from space, while more and more millimeter-range photons with higher energy are leaving the balls. The two sequences are easily distinguished.

Let us note that in an isothermal universe these differences would not exist. The (thermalized) balls would receive and emit the same radiation in each direction. It's because space is so much colder than the balls that the sequences are not identical. Now this difference in temperature is related to expansion and in particular to the introduction of disequilibrium. For a universe in equilibrium, there is no trace of time's arrow.

We can now understand the respective roles of the expanding universe and predictive horizons in the development of complexity. The hopes for freedom raised by the effect of predictive horizons would be brutally crushed by the equilibrium that would reign in a static world.[6] It's only in states of disequilibrium that the existence of predictive horizons can bring forth something new. These states are made possible by cosmic cooling.

THE WORLD SEQUENCE

TO SUM UP, we've learned that time's arrow is not tied simply to the progression from the less probable to the more probable. We understand that neither complex nor simple events in nature are exactly reversible. (To be precise, there are only complex events. "Simple" ones are falsely simple. This point is developed at greater length in Appendix B.) The reversibility of physical time is an *approximation*, valid in practice only for very short time periods.

A new filmed sequence will give us a striking demonstration of this. We first see, scattered across the screen, atoms of carbon, nitrogen, oxygen, etc., represented, according to chemical tradition, by balls of different colors.

In the second image, the atoms are grouped into small molecules. We recognize carbon dioxide, water, alcohol, and formalin, similar to what astronomers have identified in interstellar clouds.

The next image shows us more complex molecules — sugars, fats, amino acids — incorporating many dozens of atoms.

Then the double helix of DNA appears: a gigantic structure made up of several hundreds of thousands of hydrogen, carbon, nitrogen, and oxygen atoms with a little sulfur and phosphorus.

Now the screen is filled with little living cells swimming in a drop of water. Each cell is clothed in a transparent membrane and contains a nucleus of well-protected DNA molecules. Around this nucleus move myriads of giant molecules — proteins, etc.

And here are simple little organisms — sponges, medusae, sea worms. In the images that follow, we encounter, successively, fish, amphibians, reptiles, mammals, hominoids, and humans.

Did this sequence run past us in the right direction or in reverse? If time's arrow were recognizable only by its passage from the less probable to the more probable, what reply would we be likely to make? The criterion of plausibility that enabled us to identify correctly the direction of the earlier films could very well lead us into error here.

BUTTERFLIES MAKE THE SPRING

THE CRUCIAL DIFFERENCE is in the duration. The billiard sequence describes an event that lasts a few seconds. This is below the predictive horizon for the moving balls. What we have just seen encompasses fifteen billion years of evolution in the universe (well beyond all predictive horizons). It proves beyond doubt that on a cosmic scale the unfolding of time is not simply the passage from a less probable to a more probable "foreseeable."

On a scale of billions of years, physical laws enable the universe to organize itself but do not allow us to foresee the exact form this organization will take. Though it never contravenes physical laws, the future is not implicitly contained within them.

To know the laws is not, as Laplace thought, to know how to spell out a predictable future, but rather to know how a future "big" with novelty will be governed. The fate of the world is not entirely contained in the present. Without disobeying physical laws, nature can pursue its favorite activity: to innovate freely.

Each "pristine today" is made up of new events, marked by the past but not determined by it. These events influence the future and leave in their wake the possibility of other unprecedented events. Todays pave the way for more todays.

Let's return, in summing up the last three chapters, to the sight of our butterflies over a field of colza. From our new scientific perspective, what can we say about this "event"? To what degree was it predictable; to what extent did it elude predictability?

Any prediction of the future must begin with a distinction. Are we predicting a specific event or a set of events? For example, to say "It will be cold here on this day next year" is much more specific than to say "The average temperature in the northern hemisphere will be colder next winter." We can, without too much risk, put our money on the second assertion. It doesn't apply to a particular time and place, but to a collection of places and days. Individual factors, unpredictable in isolation, average out to give a predictable result.

What can we say about the appearance of complexity in the universe? Let's travel back in our minds to the beginnings of the cosmos. What could we have foreseen, given our present knowledge of the laws of physics?

The general behavior of the cosmos lends itself more easily to prediction than particular events. The formation of stars, once the temperature of the universe had lowered sufficiently, was foreseeable, but not the appearance of the sun in a particular spot at a specific moment in the life of the Milky Way. Similarly, we could predict the production of atoms and molecules through nuclear and chemical activity, but not the exact designs of frost on my windowpane on a winter morning.

We know too little about how life appeared to say if its arrival could have been predicted or not. However, several recent discoveries in astrophysics — the presence in interstellar space of molecules containing up to several dozen atoms; the detection of amino acids on certain meteorites; Urey and Miller's experiments, simulating the primitive ocean — are all sending us the same message: nature has many tricks up its sleeve. Its ability to organize material structures never fails to surprise us. In this context, it no longer seems absurd to think that life was foreseeable, even if our own existence was not at all.

As for our butterflies, since it's spring we *can predict* that they will seek to mate. We *cannot predict* the trajectory of their amorous frolicking over fields of colza. But we can foresee that this frolicking will produce other butterflies that will weave the todays of springtimes to come . . .

PART THREE

From the Point of View of Fossil Radiation

8

The Artisan of the Eighth Day

I write because I have the impression or the feeling that the world is unfinished, as though God, who created the world in six days and rested on the seventh, didn't have time to do everything. For me, the world is too small, life is too short, happiness is not happy enough. I write to complete the world, to add an eighth day to creation.

Antonine Maillet

THERE'S AN EXQUISITE DREAM that comes back to me from time to time as I wander the flowered pathways at Malicorne. In the distance, a voice is singing Mozart. The "*voi que sapete*" from *The Marriage of Figaro*. Filled with joy and willingly swept along by the meanderings of the melody, I am one with the music. My body trembles, and my happiness knows no bounds. Exalted by the dream, I'm overwhelmed by a powerful feeling of gratitude: "Thank you, Mozart. Thank you for having existed and for having composed this sublime song."

In a dreamlike mist, untouched by distance or time, I feel him there, this composer of genius, invisible but intensely present. "The earth reclaims this mortal flesh, but not the poetry," wrote Aragon on the death of Apollinaire. In this sense at least, *death is not the end of existence.*

Why is music so precious to me? Among my earliest memories, I see my mother at the piano. She is playing Beethoven's *Appassionata* sonata. My eyes just level with the keyboard, I watch her fingers as they stretch to reach difficult notes. Sound becomes a source of pleasure as, gently lifting her elbows, she sinks her hands into the white ivory. I wait for the passages where the left hand crosses over the right and continues the accompaniment in the higher registers.

This musical activity was often cut short by the arrival of my father, and the semi-secrecy with which it was conducted in my family at that time made it extremely attractive to me. "I would so much like to have a phonograph," my mother said one day. I heard myself replying, "I'll buy us one when I'm big."

We didn't have to wait that long. The first turntable, a mechanical apparatus with a horn like that for "His Master's Voice," appeared in our house when I was eight years old. I was enthralled! Along with my brothers and my sister, I made the acquaintance of the divine Mozart, his music barely audible over the surface noise from an old 78 r.p.m. With what gusto did we crank the silver-plated handle of the gramophone . . .

Our savings were all spent on records. In the basement of Archambeault's music store on Sainte-Catherine Street in Montreal, the walls were adorned with multicolored albums alphabetically arranged, introducing us to Bach, Beethoven, Berlioz, Brahms, and many others. A palace of wonders!

Thousands of hours of music to discover. An inexhaustible source of delight. Tolerant salespeople would leave us alone in the tiny, overheated booths, surrounded by so many works with poetic names: *Winter Journey*, *Prelude to the Afternoon of a Faun*, *Death and Transfiguration*. How to choose? Whenever one of us arrived home with some fresh acquisitions, it was as though new friends had entered the house.

Summers we spent at the family cottage at Bellevue on the shores of Lake Saint-Louis. There, with our cousins, we shared the music we'd discovered over the past year. Bach's cantatas melted into the warm night, along with the rustling of poplar leaves stirred by the breeze from the lake.

All through my adolescence I was intoxicated by music, but I paid little heed to musicians. In those years, Schubert was the creator of *Die Schöne Müllerin* or of the *Octet*. Nothing more. Later I came to see that, like you or me, Bach and Wagner lived human lives. I became aware that the authors of "my" music had a historical reality.

It's the moment of creation that haunts and beguiles me more than anything else. I'm extremely moved when I imagine Mozart sitting down at his table to write the *Twentieth Piano Concerto*. Or Bach preparing a new cantata for the parishioners of Leipzig.

AN IRRESISTIBLE IMPULSE

CURIOUS TO LEARN MORE about these high points of human creativity, I immersed myself in the biographies of my favorite artists, musicians, painters, and poets. Through all these varied personal histories, one theme recurred unfailingly:

an almost obsessional impulse to create, whether in suffering, as with Beethoven and Van Gogh, or with the vigor of the young Mozart; in poverty like Rembrandt or amid luxury like Rubens.

Asked about this passion that possessed them, the creators all gave the same answers: "I can't live any other way," "It's what keeps me going, it's what gives meaning to my life. It's been an obsession since I was a little child . . . "

For those who have read Winnicott (Chapter Four), these words will have a familiar ring. What's being expressed here is the will to *remake the world*. "What is in my heart," writes Beethoven, "must out, and it's for that I write." "Music must make the fire in men's minds burst forth. Whoever plumbs the meaning of my music will be free of all the misery that afflicts other men." "There is nothing more beautiful than to draw near divinity, and to shed its rays upon the human race."

This creative impulse is particularly striking in "naive art." There's a museum in Burgundy called La Fabuloserie where the works of marginal artists have been collected. They have a tremendous emotional impact. For the artists, most often uneducated, with no esthetic preconceptions, this play-related activity has frequently served as a lifeline that kept them from madness. I defy anyone to look on the "Trials of Little Mauricette" and not be touched to the quick.

THE COSMIC ROOTS OF CREATIVITY

WHAT MORE CAN WE SAY about this creative impulse, which has endowed us with splendid artistic masterpieces from every part of the globe and from every age?

My goal, during these country strolls, has been to show how the scientific knowledge of our time can shed new light on all of human culture. In the same spirit, I'd now like to tackle the enormous subject of artistic creation.

It's not a question of explaining art. The work of art exists in its own right and requires no explanation. It's pointless and sterile to think we can reduce it to a few logical or psychological elements.

My goal is different. I will attempt to discover the point at which artistic creation joined hands with cosmic evolution, as it has been portrayed to us by various sciences — physics, chemistry, biology, astronomy.

We understand, thanks to the psychoanalysts (Winnicott in particular), the crucial importance for a human being of creating an enriched and embellished reality. We'll now discover surprising parallels between artistic activity and activity in nature. *Not only do the two play the same game, but they play it under practically the same conditions.* Nature's game-playing has brought forth an almost unlimited variety of complex structures. The human being is one of the most elaborate of these. The creative artist, a product of the game, throws himself into it and pursues still further the never-ending quest for new marvels.

ART AS A PLAYGROUND

TO DEMONSTRATE THE SUBTLE RELATIONSHIP between nature and the artist, I'll compare two pictures.

A family memory will help me clarify the relationship between game-playing, indeterminacy, and artistic creation. My mother, in her old age, spent her leisure time "painting by numbers." The walls of our house were decorated with her work. On a white canvas were drawn the outlines of a landscape or a still life. Each section of the drawing had been allocated a number. This number corresponded to a paint tube of a certain color. By applying the colors indicated to the areas specified, she faithfully reproduced the painting in the illustrated catalogue.

There is no "added value" (to use the language of economists) to this activity. The instructions rule out any element of play. It is unthinkable to bring anything new to the process.

By contrast, in associating colors on a canvas, the painter tries to produce something original. Out of simple elements, he wants to create a feeling that has no precedent. He "plays," and this game results in "added value." Twenty grams of rose coral, seventeen grams of Saturn red . . . In itemizing the amounts of color applied by Picasso, we've said nothing about the beauty of the *Demoiselles d'Avignon*.

For the second picture, we imagine Johann Sebastian Bach seated at his worktable preparing a cantata for the following Sunday. His music is faithful to the musical tradition of the time. We recognize the rules and the constraints that artists were then willing to respect. Without being determined by these, the work drew upon them for its form. For his Leipzig Sundays, Bach wrote more than 340 cantatas, all different, all sublime. Absolute rules would have forced him to repeat the same cantata ad infinitum, thereby alienating his parishioners in short order.

Rules change. One did not compose in Schumann's time as in the time of Monteverdi. But the role of musical convention is always the same: to provide a framework within which and through which the work of art may be put together and take shape. On condition, once more, that the framework is supple enough to allow for diversity and variety, so that a new reality, a new range of feelings, may see the light of day and be offered as a gift to humanity.

THE GAME OF "EMERGING PROPERTIES"

THANKS TO PROGRESS IN SCIENTIFIC RESEARCH, we are now beginning to understand how nature functions. It has its tricks, its favorite recipes, its procedures, and they apply at all levels of complexity. I set out the principal elements in Chapter Six. We must briefly return to them here.

Just as Van Gogh put simple colors side by side on his canvas to bring his famous sunflowers to life, so nature juxtaposes simple elements to make new, more complex substances. These associations, in a well-defined order, result in "emerging properties." Just as the word "blue" brings to mind a color that each of its four letters does not even begin to suggest, water is a solvent, but the hydrogen and oxygen that make up its molecules are not. In the same way, the billions of neurons in our brain are not capable individually of studying astronomy.

However, the association of simple elements is not necessarily creative. Nothing new came into being on my mother's numbered canvases. A crucial ingredient is needed: a zone of freedom where "chance" and "necessity" can meet and cross-fertilize. In Chapter Five we saw how this free space was created by the expansion and cooling of the universe.

A HISTORICAL RETROSPECTIVE

THE EXTRAORDINARY DIVERSITY of forms, structures, and patterns of behavior in today's universe didn't exist fifteen billion years ago. That knowledge represents astronomy's most significant contribution to our inquiry into the nature of the world.

The photons from fossil radiation make it possible for us to put together a picture of the universe as it then existed.[1] *This picture is uniformly white.* In whatever direction we look, however fine the resolution of our instrument, we detect no "granularity," no trace of structures to come. The "gaze" of theoretical physics makes it possible to interpret this white homogeneity. The universe of that time appears as an immense, undifferentiated soup of elementary particles.

The incredible contrast between this ancient universe and our world today makes crystal-clear the direction the *history of the universe* has taken: the progressive transformation of bygone monotony into the richness of the present state.

Physics, chemistry, and biology explain the mechanics of this mutation. They make clear the nature of the forces involved and the laws that govern them. They describe the innumerable events that have changed the face of the cosmos forever.

Astronomy helps us establish a context for this transformation. Thanks to the rule of disequilibrium imposed by an expanding universe, the laws of physics only partially determine the course of events. At each stage, a wide range of "possibilities" appears. Among these, one alone will become real, without this choice being in any way foreseeable (Chapters Six and Seven dealt with this subject in detail).

It's in these zones that elementary particles may be combined into more and more complex systems, to constantly generate what is new, unpredictable, unprogrammed.

In this cosmic context, natural forces in action can become creative. That's where nature's inventiveness has its beginnings. It's the key to the stunning variety of beings and forms in our universe.

We can see it in the sky in the wide assortment of celestial bodies. There are multiform galaxies — spiral, barred, elliptical, or frankly irregular. There are stars of all colors, whose behavior is modest or extravagant. There are planets, satellites, asteroids, and comets whose chemistry is surprising.

This diversity, the stuff of beautiful picture books, is replicated on earth in the vegetable and animal kingdoms. We admire it in the countless species of wildflowers and tropical butterflies.

Variety in human beings is even more overwhelming. We are five billion on the earth, every one of us profoundly different. Each of us has a unique personality, with feelings, emotions, and a life history all our own.

The exchange of genes at the moment of conception is one of nature's "tricks" to widen evermore the range of possibilities. This formula can generate an unlimited variety of individual differences. We'll never exhaust the all-but-infinite store of identities human nature invites us to try on.

PURSUING INNOVATION

A RESULT OF FIFTEEN BILLION YEARS of evolutionary play, the human being is, to our knowledge, nature's most complex and effective creation. (By effective, I mean capable of a significant influence, for good or ill, on its surroundings.) However, we have no right to assume that we are the end point and final goal of this remarkable progression. Nothing guarantees that this beautiful child, the focus of all our narcissism, represents all that play can achieve. Today, on our planet, the human being is simply a "top of the line" product. No more, no less . . .

Humanity's potential may be seen in its accomplishments. Artistic creation is a direct extension of nature's game-playing, of which we ourselves are a product.

With the Lascaux painter or the artist of the Sistine Chapel, nature has given itself the means to pursue and perpetuate its favorite activity. It has produced a disciple who can apply nature's own "recipes" under conditions so free and effective that it can push farther than ever this game that's so dear to its heart.[2]

The artist's task is innovation. He is continuing the tradition, long established in the universe, of enriching reality. "I write to complete the world, to add an eighth day to creation," writes Antonine Maillet, Acadian poet.

WIDENING THE RANGE

BUTTERFLIES ARE NOT "gratuitously" beautiful. The splendor of their coloring is, at least in part, related to and dictated by the requirements for survival. The diversity of natural forms is not unrelated to life's major imperatives. It's necessary to eat and not be eaten. There must be children to pass on genes. These imperatives give rise to constraints that seriously limit the amount of play in biological processes.

The artist's constraints are of another sort. His freedom is greater. He enlarges the scope of nature. He can invent new tools and devise new techniques of artistic creation. The use of computers is a current example. The computer extends the range of sounds offered by traditional instruments. It creates sonorities unknown and unavailable to Bach, Mozart, and Wagner.[3]

For fifteen billion years, nature has fulfilled its creative task. Mozart wrote his music, and the cosmos is more beautiful than it was before.[4] Mozart has been dead for two centuries, but he is still with us. When his music plays, we are at one with his lyrical soul.

A MESSAGE TO THE ARTIST

WE NOW HAVE SOMETHING TO SAY about the origins of the creative impulse. Let's first point out to the inquiring artist that his way of proceeding closely resembles that of nature. He too brings together simple elements in order to generate new ones, with emergent properties. He too may be found at play in zones of freedom where the unforeseeable will flourish.

We can tell him also that his activity is part and parcel of a tradition as old as the world. It's already there in the blinding light of the primordial soup. We can trace its development through immense interstellar space, into starry furnaces, through the debris of exploded stars, and onto the surface of planetary bodies.

It's the fountainhead of cosmic diversity. To it we owe the constant emergence of new structures that succeed in producing other, still more complex structures. Step by step, it adds to cosmic complexity by bringing forth, in turn, nucleons, atomic nuclei, atoms, molecules, cells, and all organisms, including the artist who is searching his own soul.

The birth of abstract art at the beginning of the twentieth century revived the question of esthetics. Our perception of the key mover in nature's saga — game-playing with the greatest possible freedom — may be of help to us here. What is art? What are the criteria of beauty?

Nature tries everything, denies itself nothing, never censors. It's a glutton for new experiments. If they fail, it never cries over spilt milk. It moves on to the next.

The results are gauged in terms of their vitality. Not "Where does that come from?" but "What does it have to offer?" "We do not judge a tree by its roots, but by its fruit," said the Evangelist, aptly.

If, in its favorite activity, nature imposes no limits, it's hard to see why the artist should bother to do so. If there's any esthetic guidance to be gained from seeing artistic activity in terms of cosmic evolution, perhaps that is it.

In the same vein, we can take a fresh look at another question that has made a lot of ink flow. Does "the beautiful" exist in nature, or is beauty the prerogative of human artifacts alone?

In light of what we've already said, the distinction between works of nature and those of humankind becomes blurred. They belong to the same creative tradition. Beauty is first and foremost an intimate experience involving myself and the universe. To the universe I owe both the *internal* possibility of admiring the beautiful and the *external* objects of beauty there for me to admire, whatever their origin.

During the six first days, the universe brought forth the sunset that lights up the sky as I step out of the woods. On the eighth day, Mozart's melody was born, and it continues to sing in my head.

9

A Smokestack in the Countryside

To my daughter Evelyne

My contribution is to ask for a paradox to be accepted and tolerated and respected, and for it not to be resolved. By flight to split-off intellectual functioning it is possible to resolve the paradox, but the price of this is the loss of the value of the paradox itself.

— D. W. Winnicott, *Playing and Reality*

THIS PROMENADE WILL TAKE US into the Eastern Townships of Quebec, near the American border. At a turning in the path, we come upon the first buildings of a village. The avenues are lined with houses whose gardens are in flower. The municipality has received this year's prize for the best floral arrangements in the region. A little further, a factory smokestack rises stiffly into the air, belching torrents of black smoke. The trees for miles around have no more leaves. But the factory makes it possible for the villagers to earn a living.

This image came back to me recently as I was preparing a presentation for a meeting on constitutional law in Quebec. The subject of the meeting — the protection of the environment — gave me the opportunity to look at judicial activity in the light of our new vision of the world.

THE LAWS OF THE SKY AND OF ATOMS

As a child, I often heard people talking about the law. There were several lawyers in my family. These discussions held little interest for me. They seemed futile and petty. There was more to life, I thought, than problems of common walls or inheritances.

The peaceful world of the stars, the mysteries of atoms and molecules were much more attractive to me than the twists and turns and endless hair-splitting of administrative procedure. The eternal laws, universal and inviolable, that govern the movements of planets and electrons seemed infinitely more worthy of attention than human laws, changeable, regional, and, in practice, constantly violated.

Thinking about the subject in the light of modern astrophysics, I've come to revise my position. To appreciate the role of human legislation in the management of the cosmos, I first had to acknowledge a basic attribute of nature: complexity. In parallel with the "infinitely small" of atoms and the "infinitely large" of galaxies, we must also consider the "infinitely complex" of life. It's in this sphere that, despite its flaws and vicissitudes, human jurisprudence finds its justification and its pertinence.

THE FUROR OF CREATION

As with all the earlier chapters, the history of the universe as revealed to us by modern scientific research will be our point of departure for this new journey. In the first section, we'll analyze the status of legislation in the world of so-called "inert" matter. We'll then move on to the realm of the living.

The most ancient picture of the world takes us back fifteen billion years. Several times, we've had occasion to underline its importance to our discussion. It shows us a past universe that is strikingly different from what we now know. At that time, the cosmos was extremely hot, extremely dense, but above all totally disorganized. It was the "chaos" glimpsed by Homer's contemporary, the Greek poet Hesiod.

What we mean here by "chaos" is an absence of organized structures. No animals or plants, of course, but also no galaxies, no stars, not even the molecules or atoms of modern physics. The universe of that time was made up of an immense soup of "elementary particles." Free and independent, they moved about at random in this torrid space.

By contrast, let's look at today's universe. Take for example our human body with its thinking brain. Here we find the same elementary particles but in a completely different state. The billions and billions of particles of which we are made are associated with each other, ordered, and combined into an organism of amazing complexity, whose behavior is still largely beyond our understanding. Thanks to this organization we are able to perceive and to become conscious of the world around us.

The history of the universe may be seen as the story of this extraordinary metamorphosis. In *Atoms of Silence*, I showed how the various natural sciences — physics, chemistry, biochemistry, and biology — with the help of astronomy, narrated, in turn, the different chapters.

In brief, we owe the appearance on the scene of a variety of atoms and molecules to the activity of nuclear and electromagnetic forces trying to generate increasingly stable structures. This activity is governed by a number of "physical laws" that determine the course of these events.

When a star dies, the numerous atomic species produced by the nuclear force at its center are thrown off into space. There, the atoms join together, thanks this time to the electromagnetic force, to form an immense variety of molecular types. Molecular diversity is "grafted" onto atomic diversity. From graft to graft, nature is continually inventing new solutions to the problem of existence. It erects a third ladder, that of complexity.

In this chapter, I propose to examine the "legislation" required to promote this development. Our airborne butterflies from Chapter Five brought to light one of its primary features. In binding together material structures, the forces produce diversity in nature. However, the existence of these binding forces is not enough to ensure that variety and complexity will prosper. They could just as easily neutralize the complexity and lead directly to monotony. To avoid such a catastrophe, one condition must be fulfilled: the laws must not be entirely conclusive. Nothing new could ever happen if the authority of these legislative dictates were total.

As we have already seen, the expansion and cooling of the universe created "zones of freedom" that restrict and limit the absolute power of natural laws. As a result, the forces are able to enhance the complexity of the cosmos without risking its destruction at the same time. We will find that, in the world of the living, this need for flexible laws is even greater than in the realm of inert matter.

THE LAW OF THE JUNGLE

WE HAVE CONSIDERED legislation in the atomic and molecular world. We now move on to higher levels on the pyramid of complexity, where living beings, plants, and animals are to be found.

The majestic detachment of atoms and celestial bodies is a far cry from what prevails in the upper regions of complexity. Here, to cite a popular expression, "the law of the jungle" is in force. Anything goes, it's every man for himself, and might is right.[1]

Modern ethological studies confirm in no uncertain terms the immoral — or rather amoral — aspect of animal behavior. Hunger justifies all means. Nature is shrewd and effective, but ethics is not its strong suit. To guarantee the survival of the species, all strategems are permitted: lies, low blows, and deception.

A bumblebee discovers an orchid whose shape, colors, texture, even its scent match those of its female. Tricked and disappointed, it carries away fertilizing pollen, thus ensuring the flower's reproduction.

Another flower gives off an odor of carrion. Its color and texture are those of rotten meat. It pushes the art of imitation so far as to cover itself with a thin growth of hair resembling an animal's fur. A fly appears, in search of decomposing bodies. It wants to deposit its eggs where the hatched larvae will be able to find food. Taken in by the disguise, the insect leaves behind its progeny but moves on, laden with pollen, to another flower. The larvae die of hunger but the fertilization has taken place.

The cuckoo doesn't bother building a nest. It lays its eggs in the nests of other birds. To guarantee that the deceived parents will take proper care of its own offspring, it jettisons the eggs of its unsuspecting hosts.

SHORT-TERM LEGISLATION

HOWEVER, ANARCHIC EGOISM is not the only code of behavior in nature's living world. Hives and termitaries are examples of perfectly policed insect societies.

In these societies, everyone is subject to strict and immutable legislation. When an enemy approaches, the warriors move into action while, impassive, the workers carry on with their jobs. An important feature: the individual here is totally subordinate to the group. If need be, he will be sacrificed without hesitation.

Where is the civil code? How does each one know his duties and responsibilities? This is, it would appear, genetic programming so efficient it boggles our minds. However, despite the seeming perfection with which these societies function, the legislation in place is not without its problems. A closer look, as we review the situation from on high, from "the point of view of fossil radiation," reveals several potentially contentious features.

The coercive power of this legislation appears absolute. Revolutionary spirit, disobedience, and rebellion have no place. Result: a completely conservative society that goes through the same motions over and over. A timeless society that seems to rule out any possible improvement.[2] The same tough, intangible laws that guarantee harmony within the hive eliminate any possibility of progress or evolution.[3]

A second feature: this legislation is entirely focused on the good of a controlled society: a hive, a termitary, an anthill. The rest of the world exists only to be used. Ants capture colonies of insects and force them to serve their needs. They cut off their wings and shut them up in darkness in order to milk them. Everything for the ants, nothing for the enslaved insects.

These characteristics go along with another, much more serious one: *the absence of planetary consciousness*. The grasshoppers that swarm today in the Sahel have only one law: to devour as many leaves as possible. They are "ignorant" of the fact that the desertification to which they are contributing can turn against them and wipe them out. Nature has not built this knowledge into their genes.

To this type of problem, nature has "foreseen" other solutions. When there is no more vegetation, the grasshoppers are decimated by famine. The trees grow back and the grasshoppers multiply once more. As with the most vicious capitalism, nature seems to count only on the balance of existing forces to maintain life on earth.

In the long term, this balance is unstable. Nature behaves as though resources on earth were inexhaustible. The irreversible catastrophes that could come in the wake of continued environmental deterioration do not seem to be among its concerns. Nature doesn't seem to count the survival of the biosphere among its problems any more than it does individual fates. Has it forgotten to what extent our planet, created through its own efforts 4.5 billion years ago, is finite and fragile? The coming of the human race will serve as a reminder.[4]

AN UNAVOIDABLE PARADOX

THIS DISCUSSION BRINGS us up against a paradox. As a product of biological evolution, human beings *are undeniably part of nature*. However, by comparison with other living things, they function at such a high level that we are tempted to award them a special status. We can, in a certain sense, regard them as *outside nature*.

This dual situation is a source of ambiguity. The biologist Jean Dausset writes: "Nature *does not speak*, it is the human being that speaks." It would be just as defensible to say that the human being *gives a voice* to nature.

Nature, some say, has no "heart." It is indifferent to the suffering of living things. However, all through history, human beings have fought against injustice. Others have devoted their lives to compassion and humanitarian aid. Where do these praiseworthy feelings and the capacity to act on them come from if not from Mother Nature? Are not these men and women the very "heart" of nature?

This quandary can produce new insights if, instead of choosing between the two points of view, we play a different game and hang on to both of them.

To this end we'll divide the stormy relationship between man and nature into three historical periods. The two last periods, which partially overlap, will then be examined in turn from two different angles: "man in nature" and "man outside nature."

NATURE BEFORE MAN

OUR FIRST PERIOD RUNS from the beginning of the universe fifteen billion years ago to the appearance of man two million years ago.

Over time, the cosmos emerged from the initial chaos. Acted upon by physical laws, matter became more complex. Life appeared on our planet and proliferated with extraordinary diversity. The earth was endowed with millions of living species.

A contentious situation developed immediately. Fierce competition provoked innumerable conflicts. Thanks to the regulatory instincts put in place by the evolutionary process, the damage was limited. It did not threaten the planet's harmony. However, all the elements that would later darken the landscape were there in embryo.

The first living things drew their nourishment from minerals. Later came the herbivores that killed plants to live. Then the carnivores appeared, whose existence depended on the death of other animals. This "food chain" was the logical outcome of a hierarchy based on "achievement." In it, we recognize nature's constant obsession, that of bringing into the world ever more complex structures, whose behavior is increasingly elaborate.

If we agree to be naively anthropomorphic, we can ask ourselves if nature didn't feel a twinge when the first carnivore killed for food? Or doubt? Or remorse? Is that what it really wanted? Did it really mean to take that road? (To the reader put off by this personalizing of nature, use your supply of quotation marks!)

NATURE THREATENED

THE SECOND PERIOD BEGINS with the appearance of human beings in the African savannah. A series of migrations soon led them to occupy all available space on the planet.

After a gestation period of fifteen billion years, the universe, in a frenzy of inventiveness, had given birth to a prodigious "mutant." The ability to adapt and compete is what nourishes and propels biological evolution. Humanity is the product of nature's grand immorality, of its compulsion to create. And it is humanity that plays the game best, that is the champion in all classes. It meets all challenges. It adapts to all situations. It sets up house, comfortably, in all latitudes and climates. Today, it's getting ready to live in space.

With the development of science and technology, man has altered significantly the planet where he lives. He has tamed nature and transformed the countryside. Apart from the arctic landscapes, all regions have been affected to some degree by his presence.

Nothing can resist him. His influence has increased considerably with the advent of a Western civilization that, unlike traditional cultures, has no respect for nature. A large number of habitats and species have disappeared. Forests have shrunk, and parking lots have replaced vegetation. Asphalt and concrete are the outward signs of this new and threatening monotony.

Monet at Argenteuil, Seurat at Louveciennes — the names of these magical places that inspired the Impressionists have been familiar to me for years. What disappointment, what distress I felt when I first laid eyes on the banks of the Seine! Today, when I see these paintings, my pleasure is tinged with sadness and apprehension: sadness at what has been spoiled, apprehension for the future of those spaces that still remain green. What, in decades to come, will be the fate of the shaded paths at Malicorne? The danger lying in wait makes them dearer than ever to my eyes.

Before man, any instability resulting from competition and the struggle for survival was kept in check by natural safety nets. But by inventing intelligence, nature rendered these obsolete. The human being can disobey genetic commands. He can do away with himself, exterminate his neighbors, or the entire human race. He now has that capacity.

For little *Homo sapiens*, excess is everything. The sublime and the horrible go hand in hand: Wolfgang Amadeus Mozart and Adolf Hitler.

In nature, the ingenuity of the reproductive impulse knows no bounds. It has given rise to an amazing range of sexual strategies.[5] This incredible vitality, which spread living species to all corners of the globe, today threatens the earth with explosive overpopulation. It raises the specter of a world in which the slums of Cairo and Calcutta take on global proportions. The earth's surface is limited. Its ability to support life is not infinite. But nature doesn't seem to know it.

As I promised earlier, we are now going to take another look at this second stage from our two paradoxical vantage points.

Seen from the angle "man outside of nature," humanity's appearance on the scene looks like a cosmic catastrophe. Our planet is "infested" with men who seem bent on sabotaging nature's wonderful harmony. They could very well return it to its original sterility, not unlike the moon as seen at night through a pair of binoculars.

From the angle "man in nature," humanity, in this second stage, has unveiled the dark side of nature. He provides clear proof of its thoughtless behavior, like that of the sorcerer's apprentice from the tales of our childhood. He has exposed its irresponsibility and lack of planetary consciousness to the light of day. He shows how blind it is before the lethal impasse that now confronts it.

Driven by its own policy of inventiveness and its obsession to create what is ever more complex and efficient, nature seems to have locked itself into a situation that could very well turn against it. It has brought into the world a dangerous species capable of neutralizing those regulatory instincts that ensured the durability of life on earth, a species already in a position to exterminate all living things on this planet. Such are the bleak conclusions that emerge from this second stage.

NATURE'S GUARDIAN

THE THIRD PERIOD BEGAN a century or two ago. Human beings finally became conscious of the threat they posed to planetary life. It was the beginning of our ecological concern. While damage to the landscape continued and accelerated, human beings felt themselves becoming responsible for nature's future.

After a long history of aggression and brutality, after rendering extinct many plant and animal species, humanity expressed the desire to protect life. Some species that seemed doomed to extinction were saved *in extremis* . . . In Kenya, where there is a huge population of pink flamingos, it's easy to forget that just a few decades ago we thought they had disappeared forever.

Such achievements deserve to be recognized. In the context of cosmic evolution, their implications go far beyond the lives of those species that have been spared. By comparison with past human behavior, they represent hope for the future of intelligence on earth. It's on the scale of the universe as a whole that these feats assume their symbolic value.

From the vantage point of "man outside of nature," the human being is perceived, in this third stage, as he who can *save* nature. He must, like the master sorcerer of the fairytale, come to the aid of the apprentice. He must take control of an explosive situation. Humanity today has no choice but to take charge of complexity's future. It's up to us to manage nature's magnificent but irresponsible creative impulses.

From the vantage point of "man in nature," we see, at this third stage, *nature becoming conscious of the impasse it has created for itself.* It feels obliged to go beyond itself and to abandon this obsession with achievement that has been its sole preoccupation up to now. Nature now enters the world of values. Through the development of a moral sense in human beings, it is opening its eyes and becoming responsible. Man is nature's conscience.

COMPLEXITY AS A JURISDICTION

WE COME BACK NOW to this chapter's theme: *organization in the universe requires appropriate legislation*. Unlike "the infinitely large of the stars" and "the infinitely small of the atoms," "the infinitely complex of life" is not given to us once and for all. It's fragile and ever threatened. Its jurisdiction must embrace the tangled web of the human psyche. There's no going back to the Olympian simplicity and elegance of natural laws. A subtle and tolerant flexibility is needed, more in keeping with human realities.

Unlike the rigid and inescapable genetic injunctions for ants, legislation for the infinitely complex must permit those creative impulses to flourish that made complexity possible and encouraged the advent of human intelligence. Supple enough to deal with this creativity and promote it, it must accommodate and respect reality's paradoxes and contradictions.

Aware of the baleful shadows that nature has cast over itself, this legislation must channel, *without thwarting it*, the blind "ferment of cosmic vitality." We must have our cake and eat it too. We must ensure that *the ego, the group, and the planet* coexist and prosper together.

ECOLOGY AND SMOKESTACKS

WHENEVER I CAN, I take part in demonstrations and activities of the Green Party and the ecologists. In the political arena, they see farthest. They are concerned with keeping our planet habitable and preserving its beauty.

With the factory smokestack before us, belching endless columns of smoke into the countryside, we have had long and fruitful discussions. Gazing on this scene, I felt, like them, indignation and anger. However, from the vantage point of our comfortable existence and our ecological sensibilities, we easily forget the other side of the coin. The factory keeps alive the village it covers in soot. During the great medieval famines, the air was clean and the countryside unpolluted.

We must preserve both the environment and industry. We must conserve the countryside's splendor and secure our survival. As always, nothing is simple.

A CONCRETE CASE: ABORTION

WE CAN ILLUSTRATE THIS DIFFICULTY with a concrete example: the problem of abortion. What follows are the products of my own purely personal reflections on the subject.[6]

To the question "At what point does the word 'kill' apply?" we must, in my opinion, respond: as soon as male and female gametes are joined. Conception is the beginning of an ongoing process that cannot be carved up into pieces.

The mother, to justify her act, can plead "self-defense." But the child, if it could communicate, would claim its right to exist. Here we have a situation that is essentially contradictory, where no reasoning that aspires to be logical and rigorous can possibly apply.

Experience shows that most of the women concerned will abort, no matter what legislation is in place. The methods, however, tend to vary. On a kitchen table with makeshift instruments if the act is banned; in a sterile clinic with all the support of modern medicine if it is legal. This observation is enough, in my opinion, to justify a tolerant attitude toward abortion.

Far from the serene world of atoms, molecules, and planets, the wisest law will go beyond the injunctions of utopian logic to work its way into the twists and turns of a contradictory reality, while doing the least possible damage.

TO GUARANTEE AND PROMOTE A FLOURISHING COMPLEXITY

OF TOTALITARIAN REGIMES, it has been said that *everything that is not forbidden is obligatory*. Under social conditions where all-powerful legislation leaves no room for liberty, nothing new, nothing creative is possible. When official ideology claims the right to impose its directives on artists, the results are lamentable.

On the other hand, it would be practically impossible to organize one's personal life in a perfectly anarchic society. Just imagine the traffic jam that would ensue if, in a fit of libertarian enthusiasm, we were to abolish all the laws that deal with automobile traffic.

Concrete reality presents specific problems. All jurisprudence reflects a certain vision of the world. Here we maintain that only a society governed by laws that are supple and relatively unrestrictive can ensure both an organized life and flourishing creativity. This has been nature's recipe for fifteen billion years. To it we owe the richness and splendor of our cosmos.

Unlike the laws for atoms and stars, human laws are neither universal nor inviolable. They continue the work of the laws of physics, as midwives to and managers of the universe. After billions of years of evolution, they have assumed their place in the cosmic landscape as guarantors for its complexity.

10

A Steeple over Purple Beeches

The pillars of Nature's temple are alive
and sometimes yield perplexing messages;
forests of symbols between us and the shrine
remark our passage with accustomed eyes.
Charles Baudelaire[1]

MALICORNE'S STEEPLE IS VISIBLE far into the countryside. Rising over fields of grain, its gray silhouette fits well into the landscape. It tolls all the hours of day and night, and its calls to prayer bring to mind the religious fervor of times past. Today the church door, almost always locked, sends another message: the steeple has outlived tradition.

The picture of this outmoded steeple raises the question of science's stormy relationship with religion. As is our wont in these rural walks, we'll try to view religious activity in the context of cosmic evolution. What meaning might it now have, this joyous carillon in the countryside?

IS GOD DEAD?

THE HISTORIAN PLUTARCH tells the following story. During a storm, a ship's captain heard a voice crying "The great god Pan is dead . . ." On arriving in port, he carried the message to the Roman high priests, who ordered him not to tell anyone else. The message was to remain a secret, for the initiated alone. Thus it was for a long time.

In the West, scientific progress turned people against religion. Citing its "infallibility," the Church responded by hardening its philosophical positions. It tried to impose its intellectual authority by force, but more and more the "enlightened" population dropped away.

Divinity's popularity was in free fall. When Napoleon asked Laplace what part God played in his theory of creation, Laplace replied, "Sire, I had no need of this hypothesis." At the end of the nineteenth century, Nietzsche proclaimed the death of God.[2] (We will cross Nietzsche's path frequently in this chapter.) Religion was no more than a feeble bulwark against the anguish of death. It owed its survival only to popular ignorance of what really went on in the world. Simple intellectual honesty required that anyone exposed to scientific discourse had to forego any form of religious belief. Anyone with a good education would stop believing in Santa Claus. Such were the leading opinions of the time.

The videotape recorder is an instructive tool, sometimes cruelly revealing. In the course of a televised discussion on the origins of life, a novelist asked, "And what of God in all this?" General unease. "Horrors!" exclaimed a distinguished scholar. Reaching for his own book, he recited a passionate excerpt, the better to rub the unworthy lady's nose in her appalling question. The cameraman panned across the learned assembly, which had retreated into silence. When the host switched to another subject, everyone breathed more easily. The ordeal was over. We were once again in the company of well-brought-up human beings. More recently, an astrophysicist who dared admit to his religious beliefs on television was greeted with the following response from a condescending colleague: "Yours is a problem for the social sciences . . ."

SCIENCE OR RELIGION?

AS SOMEONE WHO SPENDS much of his time disseminating information about science, I am often confronted with this problem. The question periods that follow my lectures frequently turn into open forums, and many people use the opportunity to voice their philosophical positions and their concerns.

Explicit in varying degrees, Napoleon's query makes its appearance on a regular basis. Does the synthesis of sciences that I proposed in my lecture leave room for a divinity? Once the question has been asked, a murmur of relief and approval may be heard in the hall. Someone has said the words the others dared not pronounce.

First of all, let's dispose of the misunderstanding implicit in the following very popular question: "Should we believe in science or religion?" This question shows that many of our contemporaries regard science as a belief. For them, the choice must be made between two beliefs that appear to be opposed and incompatible.

Science is not a "belief." We do not expect our students to accept what we tell them without proof, as is the case with the revealed teachings of traditional religions. Our task is not simply to instruct them in scientific theory but to develop in them a critical attitude, skeptical and rigorous, without which there would be no scientific progress. Many students never learn the lesson . . . because many teachers have not learned it themselves.

DIFFERENT ROLES

SCIENCE IS NOT A RELIGIOUS BELIEF, and God is not a scientific hypothesis. What is the relationship, then, between religious and scientific teachings? Let's first look at their respective roles in the field of human endeavor.

Traditionally, the role of science was to teach human beings about the world in which they live. Astronomy, physics, chemistry, and biology explore, each in their own way, one aspect of reality. They break down nature's mechanisms into their component parts and put them together again in a coherent synthesis.

Science, as science, is not interested in the problem of "values." It is indifferent to questions of "good" and "evil." It makes no moral judgments.

It can teach us *how* to make atomic bombs or practice genetic manipulation, but it cannot reply to the moral question: *should we, or should we not* manufacture nuclear armaments or undertake a program of genetic engineering? Even the decision "to do science or not to do it" is beyond the scope of science.

Along with a world to discover, there is for each of us the fundamental question of a life to live. We're faced with the perennial problems of human existence: sickness, suffering, war, death. As societies evolve, new conflicts arise that must be dealt with. There are relationships to be fostered at every level: familial, professional, national, international. There are norms to establish and laws to devise so that life may remain livable.

In what sort of society do we want to live? What sort of existence do we want to shape for ourselves and our children? In what sort of context must the problems presented by the development of science and technology be resolved? These questions run from the proliferation of military arsenals to the construction of dams on the Loire or hydro lines on the St. Lawrence River. Must we of necessity accomplish everything that, thanks to modern technology, we have the knowledge and capacity to do? These are all questions with which we are faced, and which science cannot answer for us.

This world of values and norms has traditionally been associated with that of morals and ethics. Historically, it has been the province of a number of different religions, scattered over the surface of the globe, and it has always maintained close ties with philosophical and metaphysical systems.

In short, science is responsible for acquiring knowledge, independent of what it means to us. Religion, as the source of morality, is concerned with interpreting reality as it relates to ourselves, our surroundings, and our behavior.

Here, we come back to the idea of *evaluation* so dear to Nietzsche. "What is important," he writes, "is not so much what is true, but what helps us to live." Theories must be judged by how well they succeed in promoting life.

TERRITORIAL INCURSIONS: (A) RELIGION INVADES SCIENCE

THESE CONSIDERATIONS SEEM TO DEFINE, overall at least, the respective jurisdictions of science and religion. Now we can ask ourselves where the historical conflicts between scientists and religious figures originated.

Territorial incursions were behind a number of clashes. Often, religion invaded the territory of science. Grounding itself in revealed truth, it claimed the authority to say "how the world is made."

These clashes often coincided with new scientific discoveries. The representatives of official religions put up barriers to new ideas. They behaved as though any change in traditional teachings would call into question all Christian values, and to sacrifice one aspect of biblical cosmogony would lead to the collapse of Christianity.

At the beginning of the seventeenth century, the Vatican brought Galileo before its tribunal. The new solar system of Copernicus threatened the teachings of the Church. It was unthinkable that the Earth, site of Christ's incarnation, could not be the center of the universe. At the price of a clumsy retraction, Galileo avoided the pontifical punishment. He was nevertheless put under house arrest in Arcetri, near Florence. A few years ago, the Vatican reopened the case, and just recently it reversed its decision. Better late than never . . .

A few years earlier, Giordano Bruno was less fortunate. The dispute concerned what was still a contentious issue in astronomy, the size of the universe. "God alone is infinite," said the theologians of the time. "The God in which I believe is more powerful than yours," replied Bruno. "He created an infinite universe." In 1600, as much for his sarcasm as for his heresies, Bruno was burned at the stake in Rome.

The theory of evolution provoked another confrontation at the end of the nineteenth century. The churches rejected the animal origins of human lineage. Only human beings had a soul. If man descended from the monkey, redemption would have to be extended to all the animal kingdom, and so on.

These debates are still going on in America, where "fundamentalists" are opposed to the teaching of Darwinian evolution in the public schools. They demand the substitution of "creationist theory," derived from a literal interpretation of Genesis. God created the world in seven days. Eve was created from Adam's rib, etc. In several American states, the courts have been asked to force teachers to present Darwin's theory and "creationist theory" on an equal footing.

TERRITORIAL INCURSIONS: (B) SCIENCE INVADES RELIGION

ALONG WITH RELIGION'S INCURSIONS onto science's territory, there are many instances where science has trespassed on the traditional turf of morality and religion. Conflicts arise when, in the name of "science," the attempt is made to establish criteria for values and to impose moral codes.

Derived from Darwinian theory, notions such as "the struggle for life" and "survival of the fittest" have often been used to rationalize human behavior. The idea was to justify cutthroat competition or legitimize the warrior mentality.

There we find, just beneath the surface, the concept of a "natural" morality, according to which *it would be best to behave like nature*. What we see (or think we see) in nature must be accepted uncritically as the standard for what we ought to do. Nietzsche was scathing on the subject: "'According to nature' you want to *live*? O you noble Stoics, what deceptive words these are! Imagine a being like nature, wasteful beyond measure, indifferent beyond measure, without purposes and consideration, without mercy and justice, fertile and desolate and uncertain at the same time; imagine indifference itself as a power — how *could* you live according to this indifference?"[3]

This natural morality sometimes assumes hallucinatory forms: the extermination of the weak by Nazi eugenicists, the banning of vaccination or of cooking food by certain sects.

A recent example of territorial violation may be found in the field of ethology. This science concerns itself with animal behavior, with animals' interactions with their fellow creatures, with their social and family life. The ethologists wondered, for example, why certain individuals "agreed" not to have children. Why did "kamikaze ants" throw themselves into dangerous situations and risk their lives? In what way is such individual or social behavior beneficial to the group? How does it increase the chances of having children? How does it enable them safely to reach the age at which they may propagate?

A group of scientists, led by E. O. Wilson, offered to explain these facts within the context of "sociobiology." The celibate uncles and aunts assured the well-being of the family cell of "nephews." A small number of "kamikaze" warriors was enough to protect the ant colony against aggression from outside. But how did the individuals themselves know how they ought to behave?

According to this theory, these behavior traits were inscribed directly into the genes of the individuals of a particular species. There was, for example, the "aggression gene," the "altruism gene," and even the "homosexuality gene." This theory is far from being generally accepted by researchers.[4]

Wilson, in a controversial book, extended this thesis to include all of human behavior.[5] One school of thought, represented in France by the "New Right," wanted to read into these theses the elements of a new ethic. Much of human behavior was to be justified and legitimized by sociobiological claims and their genetic origin.

Clearly, this is a new version of "natural morality," revived under the guise of "scientific morality." Once again, deciding to "do like nature" is not a matter for science but for morality. Science, in and of itself, is incapable of telling us if we should behave like nature or not. Rather than asking "Should we behave like nature," we should ask "When is it appropriate to take nature for a model, and when is it not?" Once the question is framed in this way, the moral character of such choices becomes evident.

Such historical examples of conflicts between science and religion tie in well with this chapter's theme. Science is on home turf when it asks the question "How are things made?"; it is no longer at home when it asks "How should we live?" Religion (or all moral philosophy) is on safe ground when it deals with the problem of "how to live," but not when it tries to decide "how things are made." Territorial incursions are the principal source of their mutual conflicts.

SACRED TALES

TRADITIONALLY, EACH RELIGION founds its morality on a "sacred tale" that nonbelievers characterize as "mythology." This tale provides the framework within which life takes on meaning, and from this framework emerges the wisdom and morality specific to that religion. It's important to understand that this wisdom and morality are not necessarily called into question by any inexactitudes in the "sacred tale."

Christianity, for example, is founded on a biblical vision of the universe. Genesis provides a detailed description of this vision. Today we know the world was not created in seven days, but rather over fifteen billion years. Eve did not emerge from Adam's rib. Wars in the animal world predate by far the original sin in Eden. Noah's flood is a legend. Joshua did not render the Israelites victorious by stopping the sun in its tracks. However, the human values inherent in Christian wisdom have in no way been compromised by this.[6]

In the same way, according to the Indian Vedanta, the universe is periodically destroyed. Like the phoenix, it is then reborn from its own ashes. Such a scenario could correspond to the cosmological model of a closed universe, according to which the current phase of cooling will end some billions of years hence to be replaced by a period of gradual warming. There would then follow the destruction, by heat, of all material organization in the cosmos. Today, observations of cosmic density do not favor this model. It seems that ours is an "open" universe, with no warming to come. That's as may be. This "error" in no way calls into question the Vedantic religion. Thanks to its teachings, millions of human beings have found wisdom that has sustained them throughout their lives.[7]

DIFFERENT ROLES, BUT NOT UNRELATED

AN IMPORTANT DISTINCTION must be made here that might seem to contradict what I've just said regarding religious and scientific activity. Even if science is not a repository of values, it does have a role to play in enlightening those who make moral decisions. Before legislating, it is essential to submit all dossiers to a rigorous analysis. Social legislation that claims to rely only on principle and so ignores the behavior of populations runs a great risk of being inapplicable and unapplied.

Let's return for a moment to the issue of abortion, discussed in an earlier chapter. Preliminary sociological and psychological studies will establish the number of cases each year. They'll describe the conditions under which abortions were performed, given the guidelines in effect. This data, obtained by traditional scientific methods, will be presented to the legislators, not to take the decision out of their hands — the data make no value judgments — but so that, when moral decisions must be made, all aspects of the problem will be accessible to them.

MORALITY, SCIENCE, AND A WORLD-VIEW

THE IMPACT OF SCIENTIFIC PROGRESS on morality and religion can be even more subtle and profound.

Some scientific data alter the way we see the universe and humanity's place within it. They generate "world-views" capable of influencing philosophical and moral thought.

In Chapter Six, I described the bleak vision of the world's future that, at the beginning of the twentieth century, seemed to emerge from thermodynamic theory: a condemnation to thermal death. In his book *Religion and Science*, the mathematician Bertrand Russell drew from this certain philosophical conclusions. "There is no law of cosmic progress, but only an oscillation upward and downward, with a slow trend downward on the balance owing to the diffusion of energy. This, at least, is what science at present regards as most probable, and in our disillusioned generation it is easy to believe. From evolution, so far as our present knowledge shows, no ultimately optimistic philosophy can be validly inferred."[8]

This pessimistic world-view weighed heavily on many of the century's philosophical movements. In their turn, these movements often left their distinctive imprints on the choices and decisions of Western societies.

As we saw in "Butterflies Over a Field of Colza," we now have a different understanding of thermodynamics in the universe. Since Russell, we've discovered the profoundly innovative role of the force of gravity, thanks to which fresh sources of energy appear continually, primed to sustain the evolution of complexity.

The world-view summed up by Russell carried within it the seeds of a pessimism and disillusionment that spread far and wide. That which seems to be emerging from today's cosmology is more positive. It would encourage us to promote nature's ascent toward higher levels of complexity. We are now in a state of crisis. Our biosphere is burdened with the threats of nuclear armament and industrial pollution. In this context, our new perspective should incite us to protect, by any means, life on earth, and guarantee it the best possible conditions to flourish. We will do anything so that flower buds may continue to burst into bloom.

ARE WE REALLY ALONE?

THE PROBLEM OF LIFE'S ORIGINS is another area in which scientific knowledge can influence moral choices. Persuaded that life is an extremely improbable occurence, biologists have often affirmed that we are alone in the universe. This view could authorize Camus to say we are "strangers," and Sartre that "we are superfluous." Jacques Monod wrote, in turn: "Matter is not pregnant with life and life is not pregnant with man."

In recent decades, opinions have evolved. In the light of recent astronomical observations and laboratory experiments, the appearance of life seems much less improbable. Today many researchers think that, when the circumstances are right, it is virtually inevitable.

From the point of view of fossil radiation, far from being "strangers" we are children of the universe, the sons and daughters of stars that have generated the atoms of our bodies. Our presence is related to the most violent phenomena of the cosmos: the collapse of galactic matter, the explosions of giant stars, the dispersal of their ashes in tumultuous torrents of interstellar nebulae, the collisions of asteroids leading to the formation of the earth in the early days of the solar system. By coming to terms with humanity's place in the progressive structuring of matter in the universe, we can discover how rooted we are in the evolution of the cosmos.

This vision of the world can inform in very tangible ways the moral choices of people and societies, for it points out man's role in the vast, ongoing organization of the universe. "Strangers" in this universe would have the right to refuse any responsibility for the future of the biosphere. "Children of the cosmos," on the other hand, are intimately involved in its destiny. It is up to them to take in hand the management of our planet. It is their duty to ensure that cosmic complexity continues to flourish.

This relationship between man and the universe gives added importance to scientific knowledge and the pursuit of research programs. Not only does science tell us how the world is made, but it provides us with indispensable documentation for the preparation of dossiers that will help us make moral decisions. Furthermore, the "world-views" generated by the knowledge of a particular period influence the philosophical thought of that time and, on the rebound, what we may call "the spirit of the law." I made this point in more detail when talking about the smokestack in the countryside (Chapter Nine).

Let us, however, repeat once again that these facts, in and of themselves, carry with them no moral values. Every decision to attribute some value to them (as in the case of "natural morality") is made beyond the compass of scientific activity.

ON THE OVERESTIMATION OF CONCEPTS

JUST AS WE CAN REPROACH certain religious figures for having made God an occasion for bigotry, we can reproach certain theologians for having turned Him into a concept.

Among the sources of conflict between science and religion we must include the exaggerated confidence in the power of conceptual thought as a norm in the universe. This confidence is based on the conviction that there is such a thing as "absolute truth," which may be expressed in clear concepts and unambiguous declarations.

This conviction was held by the Greek philosophers, particularly Plato. After the birth of Christianity, it entered the realm of religion. We then witnessed a vast program of rationalization within the Christian faith. The most important effort was made by the scholastic philosophers, especially Saint Thomas Aquinas. Basing their ideas on Aristotle's philosophy, these thinkers laid the foundations of a rational Christian theology. They proved the existence of God. They described His personal attributes and the ground rules for His relationship with human beings. These declarations then became "creeds," imposed on the faithful by the supreme authority of the Church. "Heretics" were persecuted and, like Giordano Bruno, risked the stake. For a word, or two or three, war was declared. Populations that resisted were put to the sword.

For many centuries, theology and the "religious sciences"

claimed to be grounded, like mathematics, in the ethereal world of pure logic. From there, it was an easy matter to contest scientific data generated in less noble worlds — those of experimentation and observation.

Here we return to the themes of our early chapters. In hitching its wagon to logic, religious thought was condemned to the same fate. Just as I recounted the story of the empire of numbers, I could tell the tale of the empire of "religious truths": its ascent, its supremacy, and its decline.

It was left to Kant to demonstrate, at the end of the eighteenth century, the vanity of religious logic.[9] On the most fundamental questions — the existence of God and the immortality of the soul — the human mind can affirm nothing. At a blow, the whole conceptual edifice of scholasticism collapsed.

A journey similar to that which we undertook in the early chapters would lead us to seek the origins of religious practice in the "zone of freedom" so artfully revealed to us by Winnicott.

Like science and art, religious activity reconstructs the world. The many sacred tales that have been invented all over the globe have one thing in common: each in its own way, they offer the possibility of putting life's events into a coherent context. Reality is intolerable if we cannot "think it." When we can give them a "meaning," suffering and the death of dear ones can be accepted. In this sense, the traditional antireligious critique is true: religion is born from a need to neutralize the anguish of death.

The stories told by the sacred tales have little in common. We find there anything and everything. This diversity shows how helpless religions are to tell us how the world is made. A literal interpretation of all these stories would lead only to the greatest confusion.

The antireligious critique becomes questionable when it claims that religion is *nothing but* a project for moral consolation. The fact that, at the heart of religious activity, there is the need for reassurance in the face of death does not authorize us to claim that it may be *reduced* to that. I think rather that this angst is the driving force that has led human beings to seek meaning in the world.

The contradictions and scientific "errors" of the sacred tales lose their importance once we stop taking them literally and recognize that they use a symbolic language. It then becomes clear that each one, in its own way, embodies a different facet of our mysterious universe. They are neither sciences nor philosophies in the Western sense of the word, but rather wisdom in the Eastern sense. None has a monopoly on "truth," but each touches, somewhere, on hidden aspects of reality.

Religion uses a language that is appropriate to it. Words are neither vectors of precise information, as in science, nor sources of emotion, as in poetry, but rather symbols that link us with an unknown world (see Chapter Four, note 7). Take for example the word "God." This word generally refers to our intuition of a mysterious presence, inaccessible to the intellect. Kant is right to say that we can prove neither that He exists nor that He does not.

Most believers are now largely in agreement with that. Just as it has abandoned its geographical empire, the Church of Rome has given up its domination of the intellectual world.

Few believers today contest science's achievements in the name of religious "truths." Few Christians would endorse the argument that Nietzsche scornfully put in their mouths: "Science cannot be true because it denies God. Therefore it does not come from God, therefore it is not true, because God is the truth."[10]

SATURN AND THE EXPERIENCE OF THE WORLD

At MALICORNE, ON CLEAR NIGHTS, we look at the sky with a small telescope. Through its eyepiece, many friends have seen Saturn or the Andromeda galaxy for the first time. For a number of them, it was a shock. They were deeply moved.

Still, almost everyone had seen color photos of these objects, much richer and more detailed than the tiny image in my little telescope. Why the shock?

The replies are numerous and the comments vary. For example: "Through the telescope, we see it 'live,' while books only give us reproductions." Or again: "You get a sense of depth in the sky that a picture can't provide. The sky has three dimensions, whereas the page has only two." Or yet again: "We're always a bit suspicious of what's been published."

None of these answers is really convincing. To the first we are tempted to reply that, for the eye, the telescopic image, like that of the book, is seen in the same way, as a result of photons striking the retina. To the second, that at the distance of Andromeda and Saturn, our sense of three dimensions is utterly useless. To the third, that any distrust of NASA or the great astronomical observatories is hardly justified today. Who would suspect them of falsifying their documents?

I think the true source of the emotion is to be found elsewhere. Perhaps it has something to do with the mysterious and inexplicable phenomenon that we might characterize as our intimate rapport with reality. I believe that religious experience is of the same order.

GOD IS NOT WHAT HE WAS

THAT PEOPLE WHO ATTEND LECTURES on astronomy have a compulsion to bring up the subject of religion seems highly significant to me. If I had to summarize in a few words the impression I'm left with, I'd say that God, far from being dead, is very much alive to our fellow citizens. But he's not what he once was.

His status has changed profoundly. His kingdom is no longer that of the intellect. He is no longer "the Truth." "What is the truth?" asked Pontius Pilate two thousand years ago. This insidious question has prevailed over Rome's claims to infallibility.

Now God is to be found elsewhere, entirely. We encounter Him through questioning, and no longer in the realm of certitude. He takes his place in the inner travelings of each one among us. He is the hidden itinerary for this voyage that lasts as long as we exist. He is inseparable from our doubts and the questions we ask about the deeper meaning of things.

He is linked to the quiet conviction that, beyond what we are allowed to see, there is "something" in which we are profoundly, vitally, existentially implicated. A "something" to which nothing is foreign — not the woodland violet, not Feynman's diagrams, not the Nazi extermination camps, not Mozart's *Requiem*.

Far off, rising over the tiers of foliage, I catch sight once again of Malicorne's church steeple. Though largely abandoned, this modest architecture continues to stand for our deepest feelings before the unfathomable mystery that is the universe.

APPENDICES

A

Maximal Entropy and Real Entropy

IN AN EXPANDING UNIVERSE, real entropy increases *less rapidly* than maximal entropy. We can demonstrate this in the following way.

It is important to define correctly the spatial volume we are going to consider. Let's first of all review the notion of a "sphere of causality." This is a sphere whose radius approximates the product of the speed of light and the age of the universe. No contact beyond this sphere is possible. We could also say that the volume of this sphere is limited by the "cosmological horizon." The maximum quantity of energy available at a given moment is that which is contained in this sphere. This, therefore, is the volume we must choose for our evaluation.

We may obtain the maximal entropy in this volume by transforming all the energy found there into a single black hole whose mass would then be equivalent to this energy (using the Einsteinian equivalence). The sphere of causality expands as time passes, and the quantity of energy it contains grows proportionately. Maximal entropy is therefore not a fixed and unchanging quantity. It increases with the rhythm of expansion. It can be shown that it grows in proportion to the square of time (time to the power 2).

The most important contribution to real entropy in the universe comes from fossil radiation. It also increases along with the sphere of causality, but more slowly than maximal entropy (proportionate to the power *three-halves* [3⁄2]) of time, when the universe is dominated by the radiation's density, and to the power of *one* when matter dominates.

The result is that the relationship between maximal entropy and real entropy increases with the square root of time (to the power of ½) when radiation dominates, and linearly with time when matter dominates. Thus we conclude that in an expanding universe, and for as long as this expansion will last, the state of maximal entropy will never be reached.

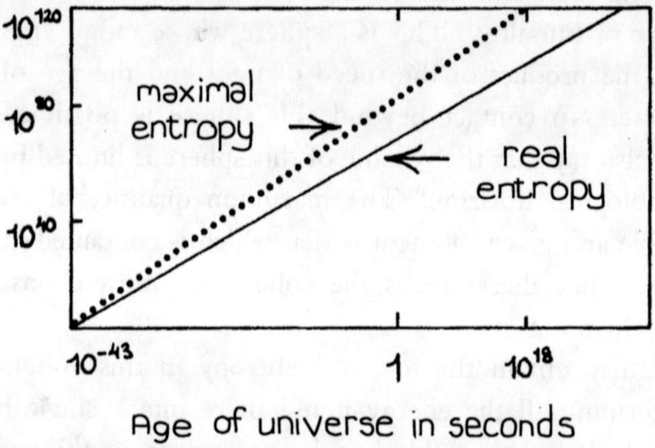

The growth of entropy in the cosmos. *On the horizontal axis, we have put the time in seconds (logarithmic units), beginning with Planck's time (10^{-43} seconds). On the vertical axis, maximal entropy (dotted line) and real entropy. The difference, always increasing, gives the measure of cosmic* information. *It is the source of organization in the universe.*

B

Temporal Reversibility of Physical Phenomena: *In Vitro* and *In Vivo*

A PARADOX, OFTEN CITED IN PHYSICS, concerns the reversibility of time in elementary physical processes, contrasted with the irreversibility of events in real life: an ice cube melting in a glass of warm whiskey, the shattering of a crystal goblet as it strikes the ground.

This irreversibility is traditionally linked to the notions of a *probable situation* and an *improbable situation*. The key word is *plausible*. It's not impossible, only excessively improbable that, in a glass of cold whiskey, an ice cube will spontaneously form. It's infinitely more probable that we'll see an ice cube melt. We therefore choose the first sequence because it is plausible, whereas the same cannot be said for the second.

Let's take as a model for an elementary reaction in physics the elastic scattering of an electron by another electron. In the initial stage the two electrons approach each other; in the final stage they move apart. The equations that describe this reaction are the same as those that would describe the situation if time were reversed. The initial stage of the first reaction becomes the final stage of the second and vice versa.

It is often said that time's arrow is not inscribed in this simple event: past and future are indiscernible. This event is in a state of reversibility with regard to time (temporal reversibility).

IN VITRO AND *IN VIVO*

BUT LET'S TAKE a closer look. We've willfully neglected all the other interactions these electrons have with the rest of the universe. We've mentally isolated them. We've conducted what biologists call an experiment *in vitro.*

The situation *in vivo* is different. The universe is not empty; space is full of radiation. In the first place, there's cosmological radiation at 2.7 K (fossil radiation), made up of photons of energy of about a millielectron-volt. The Big Bang theory also posits the existence of "fossil radiation" from neutrinos and gravitons. To that we must add the energy from the so-called "quantum vacuum," which takes the form of the creation and annihilation of particles of matter and antimatter. In other words, reactions involving two bodies are theoretical fictions that bear no relation to reality.

The intensity of these fields is itself linked to the density of cosmic matter. *And this density lessens in the course of time.* Here is the factor whose precise role we must define, as it relates to time's arrow.

We might be tempted to take the following route. Let's consider the fact that, in the context of *in vivo*, our test-electrons are subject to interactions with supplementary particles. However short the trajectory we have in mind, there is a not insignificant probability that the electron will be affected and its path altered. This alteration is not the same when the sequence is reversed, because in the "true" case the density *decreases* during the journey, while in the "reversed" sequence it *grows*.

It would, however, be inexact to conclude that we have discovered a way to identify the direction of time's arrow. All we know is that one of the sequences takes place in a universe that is cooling, while the other takes place in a universe that is growing warmer. We have simply learned to relate the trajectory of a particle to the dynamics of the universe.

EXPANSION AND IRREVERSIBILITY

EXPANSION, INSOFAR AS it's a dynamic phenomenon acting on the totality of galaxies, is a reversible phenomenon. It does not bear the mark of time's arrow. It creates no entropy. Its temporal reversal, contraction, is also described by the reversible equations of the cosmological model.

Let us suppose, to clarify our thinking, that our two test-particles are two high-energy electrons of cosmic radiation. The interaction of these particles with millimeter-range photons of fossil radiation triggers the emission of a sequence of ultraviolet photons (*bremsstrahlung*) that progressively slows their movement. In the "true" sequence, the electrons absorb millimeter-range photons and emit the UV photons that cause their deceleration. In the "reverse" sequence, they absorb UV photons coming from the outside world — which accelerate them — while emitting millimeter-range photons. The idea of a sequence of UV photons arriving out of the cosmos just at the right point to accelerate the electrons on their trajectory is sufficiently improbable to reveal time's arrow.

But what is the connection between the expansion of the universe and the deceleration of electrons of cosmic radiation along their trajectory? The thing to remember here is that the energy of these electrons is much higher than that of the fossil radiation through which they move. Matter in the universe is not in a state of thermal equilibrium where, overall, all the particles would have approximately the same energy and where accelerating collisions would be as plausible as decelerating collisions.

This was not the situation in the very distant past — extremely dense and extremely hot — of our universe. At that time, matter was in thermal and reactive equilibrium. Reactions of the type A + B = C + D induced by different interactions were in equilibrium with the reverse reactions C + D = A + B (that is, just as frequent).

If expansion as a dynamic phenomenon is reversible, nevertheless this same expansion is, indirectly, responsible for the regimes of thermal and reactive disequilibrium that exist in our modern universe. One of the characteristics of these regimes is the fact that the words "plausible" and "implausible" are current there. Certain reactions are probable, whereas the reverse are improbable. It's thanks to the existence of these regimes that we can identify time's arrow (by the deceleration of the electrons of cosmic radiation, for example).

How did expansion lead to the introduction of these regimes of disequilibrium? In general terms, it results from the fact that the rhythm of expansion and cooling of cosmic matter is too rapid to allow the different interactions to sustain their state of primordial equilibrium. If, today, there are particles whose energy is greater than the average energy of photons of fossil radiation (the electrons of cosmic radiation, for example), it's thanks to the introduction of these states of disequilibrium, and thus thanks to expansion.

To sum up, although expansion is a reversible phenomenon, it is indirectly responsible for the introduction of these regimes of disequilibrium that inscribe time's arrow in *all* of reality's phenomena, once we agree to study them in their real context.

Let us return now to our classroom case: the electrons are *in vitro*, that is to say arbitrarily isolated from all interactions other than with themselves. Moreover, only electromagnetic interactions are allowed.

In practice, it amounts to putting the electrons into a hypothetical enclosure that would be perfectly empty and would provide complete isolation from all interactions. But if such an enclosure were practicable, our study of the physical situation would lead us to redefine the system studied. This would consist now of the sum of two electrons and the enclosure itself. Now this enclosure interacts with the outside world, and thus would be subject to the disequilibriums referred to above. And so we're thrown back on our earlier example. There's no more reversibility *in vitro* than *in vivo*.

C

Quantum Fluctuations and Pythagorean Mythologies

The idea has sometimes been proposed that our universe may have been born from a "quantum fluctuation," triggering or accompanying a "rupture in space-time" out of which everything we observe — galaxies, stars, planets, atoms, molecules, and also living organisms, plants and animals, including of course ourselves with our inquiring minds — would have evolved. Nothing is more tempting than to equate this fluctuation with the transition from "nothingness" to being, and to claim to have answered Leibniz's famous question as to why there is "something" rather than "nothing."

Before going on, we should note that these are the same quantum fluctuations that were first formulated in the context of atomic physics between 1920 and 1930, by Werner Heisenberg in particular. We may note as well that quantum theory imposes limits on the definition of certain properties of matter. In particular, in the case of an unstable system, that is to say one that exists only for a certain time (a radioactive atom such as uranium, for example), it asserts the impossibility of *defining* (and therefore of *determining*) with extreme precision *both* the system's energy and the duration of its existence. Any improvement in accuracy of one of these two parameters will be at the expense of one's ability to determine precisely the other parameter.

Let us hasten to add that this curious property, called "quantum indetermination," is not an artifact that can easily be set aside. It's part and parcel of the theory. To reject it would be to reject the theory as a whole. Given the extraordinary effectiveness of the theory, that would be a shame.

In physics, a "vacuum" is a volume of space in which there is no matter and no energy. Quantum physics makes an astonishing assertion: such a volume cannot remain empty, for a finite period, without violating this principle of indetermination. Otherwise we would be able to exactly determine both the quantity of energy in the physical system this volume represents — as it happens, a zero quantity — and the length of time the system will exist.

We observe, indeed, that there are "created," spontaneously, energies called "vacuum energies." If the reader is astonished, then so were the physicists, long before. But they came to accept it . . . Nature sometimes behaves in strange ways. We have to take her as she is . . .

These energies generally appear as pairs of particles, for example, electrons and positrons. The energy contained in such a pair is about a million electron-volts. After a very short time, the pair annihilates, and the system returns to its initial state. This phenomenon, called "quantum fluctuation," repeats itself numerous times. This "humming" affects, in particular, atoms introduced into the system (the Lamb effect). Laboratory observations confirm with great precision the reality of these phenomena.

The creation of new particles is not limited to pairs of

electrons and positrons. Much more massive pairs of particles may also create and annihilate themselves. The principle of indetermination implies that the more massive the particles, the shorter their life-span. The limit is provided by Planck's mass (about 20 micrograms).

This amazing phenomenon naturally raises the possibility of the entire universe's being born from a quantum fluctuation. In the context of quantum cosmology, this idea takes the form of a model universe incorporating all the properties of matter.

We might, however, ask ourselves if such an approach has anything to do with Leibniz's question. The approach consists of our imagining, *in the first instance*, an empty universe into which quantum fluctuations might, *in the second instance*, have injected a quantity of appropriate matter. In other words, the "nothing" out of which one wants the "something" to emerge is a universe that, while empty, already fully obeys the laws of physics, for it's only within the context of these laws that quantum fluctuations can occur.

In other words, and still in the spirit of Leibniz's question, this view implies that physical laws are pre-existent — in an ontological sense — to matter, time, and space, and may be thought of as "nothing," while matter alone deserves to be called "something." We are right back to the Pythagorean mythology.

We must remember here that, in contrast to the "vacuum" in physics, the "nothingness" of metaphysics must be truly empty. That is to say no time, no space, and above all no prior rule of physical laws. Nothing is nothing . . .

This discussion calls to mind Saint Augustine's Eleventh

Confession. Asked the question "What did God do before the creation of the world?" he first, to show his disapproval, voiced a traditional fifth-century response: He made Hell ready for those who tried to fathom divine mysteries. Saint Augustine was an enlightened spirit who encouraged inquiry. His own reply was that the question had no meaning. It supposed that time existed before the creation of the universe, the time during which one wondered what God was doing. Time, said Augustine, with an Einsteinian flair, may not be disassociated from the material universe with which it appeared at the moment of creation.

The same sort of consideration applies to the models for "baby universes" that have caused much discussion in recent years. There, inspired by the properties of black holes, the properties of extremely intense gravitational fields have been used to generate new universes, completely independent from our own. But always in a context where a formative physics exists prior to the appearance of all the newborn.

Notes

Chapter 1: The Empire of Numbers

1. On ancient science, I recommend *Pythagore Superstar* by Luciano de Crescenzo (J. C. Lattès, 1985), as well as *Les Penseurs grecs avant Socrate* (Flammarion, 1964), and *The Mind of God*, P. C. W. Davies (Simon and Schuster, 1992).
2. This is curiously reminiscent of the work of modern computer scientists, who evaluate the number of "pixels" needed to record and transmit a document.
3. Roger Penrose, an English cosmologist interviewed in *Scientific American* (November 1989, p. 32), confesses "to agreeing with Plato that the truth is embodied in mathematics and exists 'out there,' independent of the physical world and even of human thought. Scientists," he adds, "do not invent the truth — they discover it." See *The Emperor's New Mind* (Oxford University Press, 1989).

   In a similar vein, Louis de Broglic writes: "The inventor has, suddenly, the distinct impression that those ideas that he has just now brought to light, insofar as they are correct, already existed before they were ever conceived in the human brain," *Le Continu et le Discontinu en physique moderne* (Albin Michel, 1941), p. 81.

4. Bertrand Russell, *The Problems of Philosophy*, p. 156, and *Mysticism and Logic* (London, 1919), p. 241.
5. Descartes, *Meditations on First Philosophy, from Selected Philosophical Writings*, translated by John Cottingham, Robert Stoothoff, and Dugald Murdoch (Cambridge University Press, 1988). Here is the exact text with some supplementary paragraphs:

> . . . there are countless particular features regarding shape, number, motion and so on, which I perceive when I give them my attention. And the truth of these matters is so open and so much in harmony with my nature, that on first discovering them it seems that I am not so much learning something new as remembering what I knew before; or it seems like *noticing for the first time things which were long present within me* although I had never turned my mental gaze on them before.
>
> But I think the most important consideration at this point is that I find within me countless ideas of things which *even though they may not exist anywhere outside me* still cannot be called nothing; for although in a sense they can be thought of at will, they are not my invention but have their own true and immutable natures. When, for example, I imagine a triangle, even if perhaps no such figure exists, or has ever existed, anywhere outside my thought, there is still a determinate nature, or essence, or form of the triangle which is immutable and eternal, *and not invented by me* or dependent on my mind. This is clear from the fact that various properties can be demonstrated of the triangle, for example that its three angles equal two right angles, that its greatest side subtends its greatest angle, and the like; and since these

properties are ones which I now clearly recognize whether I want to or not, even if I never thought of them at all when I previously imagined the triangle, it follows that they cannot have been invented by me.

It would be beside the point for me to say that since I have from time to time seen bodies of triangular shape, the idea of the triangle may have come to me from external things by means of the sense organs. For I can think up countless other shapes which there can be no suspicion of my ever having encountered through the senses . . .

Chapter 2: The Decline of the Empire

1. Linguists will tell me that the word "invent" comes from the Latin *invenere*, which means "to find," therefore "to discover." Here I give it its more traditional meaning, akin to "to create."
2. Here are the thoughts of some great thinkers and mathematicians on the subject, as cited in Morris Kline, *Mathematics: The Loss of Certainty* (Oxford University Press, 1980), p. 324.

   Dedekind: "We [mathematicians] possess the power to create."

   Weierstrass: "The true mathematician is a poet."

   Wittgenstein: "The mathematician is an inventor and not a discoverer."

   Morris Kline himself: "In favor of their position is the fact that such elementary concepts as irrational numbers and negative numbers are neither deductions from empirical findings nor entities obviously existing in some external world."
3. Over the last few years, a new field of mathematics has grown up around "chaos theory." We'll have occasion to talk about it

in the second part of the book. It's become clear that the number of mathematical systems involved in the description of reality is much larger than was believed up to this point. In particular, new geometries, called "fractals," have proved extremely useful in describing phenomena thought irreducible to mathematical formulation. The word "fractal" indicates that these are not geometries of two or three dimensions but of fractional numbers of dimensions, for example, 1.63.

4. Bertrand Russell, *Portraits from Memory* (1958).
5. Stephen Hawking, *A Brief History of Time* (Bantam, 1988).
6. Of course, such temperatures are beyond the capabilities of current scientific technique. It's doubtful that we'll ever be able to reproduce them in the laboratory. How do we know these difficulties exist? Because of unacceptable contradictions in the mathematical formulations themselves. For example, probability calculations that give infinite results.
7. R. Pirsig, *Zen and the Art of Motorcycle Maintenance* (William Morrow and Company, 1974). We might better formulate the question differently: "Did one pebble set beside another pebble make two pebbles in the age of the dinosaurs?" Pirsig's version assumes the existence of "two." Even the existence of "unity" at this time is questionable.
8. I found this citation in Rémy Chauvin's book, *Dieu des fourmis, Dieu des étoiles* (Pré-aux-Clercs, 1988).
9. We may illustrate the situation further by asking our duo "brain" and "thought" to perform their clock and clockmaker number.

   In the first instance, "thought" inquires about itself and discovers that it has been produced by "brain," a sort of machine to manufacture thought. There is no thought without brain, just as there is no clock without a clockmaker. Thank you, Voltaire.

But "brain" tells us that every machine corresponds more or less to a blueprint whose devising presupposes the prior workings of "thought."

Wherefore the duality: which came first? Brain or thought? The interplay between these two elements is richer in its dialectic than the duality of the clock and the clockmaker, because the clockmaker is a dead end.

But is it possible to conceive of what comes first? And if so, can it really be first if the concepts that describe it precede it, not to mention thought itself . . . ?

10. Piaget, *Psychology and Epistemology*, translated by Arnold Rosin (Grossman Publishers, 1971), p.118. I would contest, in passing, the reductionist use of the negative form: "the universe is known to man *only* through logic . . ." On this subject, I prefer the more holistic vision of a Saint-John Perse, for example.
11. Lynn Margulis and Dorion Sagan, *Microcosmos: Four Billion Years of Evolution from our Microbial Ancestors* (Simon and Schuster, Inc., 1991).

Chapter 3: Construction

1. On this topic, read Chapter 10 of *Physics and Beyond* by Werner Heisenberg (Harper Torchbooks, 1972).
2. We must not forget that birds are descended from reptiles, and are therefore cousin to dinosaurs. We might, in jest, reply to Pirsig that perhaps, after all, dinosaurs already knew that 2 + 2 = 4 . . .

   Even more amazing is the musical knowledge of birds. Apart from human beings, they alone in the animal world know how to sing. Certain researchers go further and claim that thrushes, for example, possess a musical language similar to our own. Szoke, a German ethologist, claims to have discovered, analyzing their song, the presence of fifths and octaves, the musical intervals on which Pythagorean thought is based. This affirmation, it must be said, has not up to now met with general approval on the part of ethologists. Too bad — Pythagoras would have been delighted! In any case, these birds are certainly not aware that, to reach the octave, they must reduce the length of their vocal chords by half. Perhaps it's better thus, if we're to trust in the wisdom of this Chinese fable:

> A centipede was happy quite,
> Until a frog in fun
> Said, "Pray, which leg comes after which?"
> This raised her mind to such a pitch,
> She lay distracted in a ditch
> Considering how to run.

3. Deleuze has written, in this connection: "Empirical psychology, it turns out, is based on a transcendental topology. It is tempting to reply that this transcendental topology itself is a construction of the human spirit, and in this sense, in its turn, is a product of empirical psychology . . ."
4. Claude Lévi-Strauss, *The Naked Man*, translated by John and Doreen Weightman (Harper and Row, 1981), p. 688.
5. What is going on at this point, within the brain? Neurophysiologists have taught us the importance, for brain function, of neuronal connections (we will return to this later). Throughout our life, "cabling" between the cells in our cortex is connected and disconnected. It's possible that the flash of discovery may be associated with the appearance of new connections. But

how does cabling relate to our acquiring a new idea? How can the connection of a cable help us to understand that no water is left in the first glass after we have just filled the second, *therefore* the glasses have the same volume?

6. Jean Claude Bringuier, *Conversations libres avec Jean Piaget* (Laffont, 1976). English language edition: *Conversations with Jean Piaget*, by Jean-Claude Bringuier, translated by Basia Miller Gulati (University of Chicago Press, 1980).
7. Sigmund Freud, *Leonardo da Vinci: A Study in Psychosexuality*, and Arno Stern, *La Mémoire Organique*.
8. Let us note in passing the somewhat paradoxical situation in which this research, inevitably, finds itself. Scientists must use their own brains to try to understand how the brain works. That mischievous duo, "brain" and "thought," appear once again over the horizon. But the ambivalence of this situation has never prevented men and women of science from setting to work. Fortunately. In their own way, modern researchers are like the daredevil explorers of past centuries who ventured forth, undaunted by the cannibals' black cauldrons from out of our children's books.
9. For those who want to read further, I recommend *The Amazing Brain*, by Robert Ornstein and Richard Thompson (Houghton Mifflin, 1986). As much for its simple text as for the magnificent illustrations by David Macauley, this book is a pedagogical gem.
10. Through connections called dendrites, each neuron receives impulses from other neurons. These impulses may provoke it to "discharge," in turn, toward other neurons by means of a special connection called an "axone." Into this axone it sends an electrical impulse. At the end of the emitting axone there is a little

bulge called a synapse. The synapse does not quite make contact with the dendrites (receptors) of the other neurons. Between these elements there is a gap called a "synaptic space." This gap is crossed not by the electrical impulse but by molecules within the synapse that the impulse activates when it arrives.

11. Recent studies seem to favor the role of dendritic connections in the creation and preservation of a memory (*Scientific American*, July 1989). The effect of learning a new reflex would be to lower the threshold of excitability for a set of neuronal connections. This is a new version of an old idea: memories inscribed in the brain like gullies on a cliff face. Runoff water, working its way down, digs deeper and deeper channels.
12. We find in Lévi-Strauss a similar ideology when it comes to the famous "empty boxes." To him, they are the embodiment of an immense clockworks on the scale of humanity as a whole. Structuralism might be said to have broken down, into units of "empty boxes," the mechanism of this clockworks. On a more philosophical level — although he denies that he philosophizes — he sees evidence for a kind of deterministic view of human behavior, not easy to reconcile with the idea of freedom and autonomy for the individual.

He writes further: "The ego is not only abhorrent, there is no room between *us* and *nothingness*." In this perspective, the goal and effect of his ethnological analysis would be to demonstrate that the "ego" is an illusion, that in fact we are only expressions of a collective whole, and that "I," finally, does not exist. (This idea crops up as well in Foucault, in Freud, and, much earlier, in all Zen philosophy.)

Lévi-Strauss went even further in interpreting his research, during a literary exchange with the philosopher Jean-Paul Sartre. Replying to an accusation that he was an "aesthete," he replied: "We accept to be characterized as an aesthete insofar as we believe that the ultimate goal of science is not to form man, but to dissolve him."

But what does the word "dissolve" mean, for Lévi-Strauss? Elsewhere, he reminds us that *dissolve* is related to *solvent.* This is the language of chemistry. We think of water dissolving solid bodies — thereby taking away their identity — into a large number of molecules. Lévi-Strauss's idea would seem to be the following: under the eye of science, the "human ego" would dissolve into its constituent elements.

A more detailed discussion of this issue may be found in *Comprendre Lévi-Strauss*, by Jean Baptiste Fages (Privat, 1972).

13. Jean-Pierre Changeux, *Neuronal Man* (Pantheon Books, 1985).
14. According to Minsky (cited in *Les Vrais Penseurs de notre temps* by Guy Sorman [Fayard, 1989]) "There can be no opposition between matter and mind because it is matter that produces mind in conformity with laws that are yet to be discovered." To complete this vicious circle, one need only add: "Matter produces mind which produces the laws that organize matter in such a way that it may produce mind . . ."

Chapter 4: The Zones of Mental Construction

1. In *Structural Anthropology*, Lévi-Strauss writes: "The unconscious is always empty; or more exactly it is as removed from images as the stomach from the food that passes through it."

   On his disagreement with Jung, he says: "What proves the inadequacy of the Jungian approach is that myths, to all appearances creative, spontaneous, bountiful, nevertheless resemble each other from one end of the earth to the other." (Fages.) I've never been able to understand how the one proves the other . . .

2. "Conscience du corps," *Cahiers jungiens de psychanalyse*, 55 (1987), p. 1.
3. I had the good fortune to interview Françoise Dolto not long before her death. Her development and her work placed her exactly at the crossroads that interested me. As a children's therapist, she found herself often at that stage in maternal relations where the child does or does not begin to "learn" and to "create." My intention was to extract from these interviews passages that could lead us a bit further along in our investigation of the beginnings of thought. In fact, that fascinating meeting never gave me anything to write. What I witnessed was more of a spectacle. Everything was concrete, rooted in case histories. Numerous variations on a theme: the crucial role of the body, of the relationship with the mother, of speech, as necessary conditions for the development of thought.

   All these inquiries bring us back inexorably to the mysterious continent the psychoanalysts call the "unconscious," where the mind plunges into the body, to emerge again, as long as an attentive mother is present at the child's bedside.
4. "Reality cannot be reduced to language. It must be approached with a certain dread, mixed with pride, like matter in fusion and in flight, existential magma, everlasting chaos (and by definition indifferent to Logos). Hold to it even if no word can do it justice and no writing contain it.

   "Any writer worthy of the name works with the illusion and

the megalomaniacal intention of grasping reality, and in fact succeeds only in defining a relationship to it and taking a precise position."

5. "The Winnicottian concept of transitional phenomena is crucial to an understanding of how the internal world is related to the external world; the transitional object is that object in the external world that contains or includes intrapsychic content or fantasies. Cultural objects . . . are understood as transitional shared objects." Richard Koenisberg, *Psychanalyse dans la civilisation*, 1 (October 1989).
6. "When everything is going well, the mother has within herself the resources which enable her to spare her infant the horrors of that 'nameless terror' accompanying separation.

   "Positive early experiences prepare the way for . . . the transitional object (teddy bear, diaper, handkerchief). The normal infant will be able to distinguish this object from its body, and . . . will not close the door completely to the 'not-me.'" Frances Tustin, *Autisme et Psychose de l'enfant* (Le Seuil, 1977), quoted in *Psychanalyse dans la civilisation*, 1 (October 1989).
7. The Danish physicist Niels Bohr proposed a similar relationship between the language of science and that of religion: "To put it in an exaggerated way: in religion we renounce the wish to give words an unequivocal meaning from the outset, while in science we start with the hope — or, if you like, the illusion — that one day it may be possible to do just that." Quoted in *Physics and Beyond*, by Werner Heisenberg, (Harper Torchbooks, 1972).
8. Charles Baudelaire, *Les Fleurs du Mal*, translated by Richard Howard (David R. Godine, 1982).

9. "All the magnificent achievements of mathematical and physical science . . . proceed from our indomitable desire to cast the world into a more rational shape in our minds than the shape into which it is thrown there by the crude order of our experience." William James, *Essays in Pragmatism* (Hafner, 1969). Perhaps. But this leaves completely open the question of why this "recasting" is so much more effective, so much more reliable than the "crude" form registered by our sensory experience.
10. A passage from Malebranche, cited in *Physics and Beyond*, by Werner Heisenberg, *op. cit*: "The same tendencies that provide for visible order in the world, for the existence of chemical elements and their properties, for the formation of crystals, for the creation of life and everything else, may also have been at work in the creation of man's mind. It is these tendencies which cause ideas to correspond to things and which ensure the articulation of concepts. They are responsible for all those real structures that only become split into an objective factor — the thing — and a subjective factor — the idea — when we contemplate them from our human standpoint, when we fix them in our thoughts."
11. I think we can speak of *chaos*, in this context, much as we speak of *nothingness* in relation to *being*. Bergson writes that "nothingness is an idea destructive of itself, a pseudo-idea, just a word." Sartre writes, "It is not being which emerges from nothingness, it is nothingness that can be conceived — insofar as it is conceived — only in terms of being." These two statements seem to me to apply just as well to the word *chaos* in relation to the word *order*.

12. "There is often more mystery than we suppose in the simple fact that a little science is possible," Louis de Broglie, *Le Continu et le Discontinu en physique moderne* (Albin Michel, 1941), p. 81.

Chapter 5: Butterflies and the Laws of Physics

1. Translated by Patricia Terry and Maurice Z. Shroder, from Stéphane Mallarmé, *Selected Poetry and Prose*, edited by Mary Ann Caws (New Directions, 1982).
2. In this analogy I have given "today" two different roles: it is at once the *swan*, that *used to be* imprisoned in the ice of determinism, and the *liberating blow of the wing* from chaos theory.
3. "We ought then to regard the present state of the universe as the effect of its anterior state and as the cause of the one which is to follow. Given for one instant an intelligence which could comprehend all the forces by which nature is animated and the respective situation of the beings who compose it — an intelligence sufficiently vast to submit these data to analysis — it would embrace in the same formula the movements of the greatest bodies in the universe and those of the lightest atom; for it, nothing would be uncertain and the future, as the past, would be present to its eyes. The human mind offers, in the perfection which it has been able to give astronomy, a feeble idea of this intelligence. Its discoveries in mechanics and geometry, added to that of universal gravity, have enabled it to comprehend in the same analytical expressions the past and future states of the system of the world." Laplace, *A Philosophical Essay on Probabilities*, translated by Frederick Wilson

Truscott and Frederick Lincoln Emory (Dover Publications, 1951), p. 4.

4. Einstein was one of the last holdouts for the necessity pole. His belief in an absolute determinism profoundly influenced his world-view, going far beyond the sphere of his professional activity. On the death of his friend Besso, he wrote the following lines: "Now, although he has departed this strange world a little before me, that is nothing, because we who believe in physics *know that the distinction between the past, present and future is only a stubbornly held illusion.*" (Cited by Fang Li Zhi in *Creation of the Universe* [World Scientific, 1989].)

   I find this text fascinating and profoundly moving. I have difficulty imagining that such a belief can be a source of comfort on the occasion of a friend's death.

5. Where do these cosmic rays come from? Stars are continually shedding matter into space. A tiny fraction of the atoms thrown from their surfaces is accelerated to energies of billions of elecron-volts. Solar storms and eruptions inject rapid particles into the planetary system.

   These particles spread throughout the galaxy. Under the influence of interstellar magnetic fields, they continually change direction, losing all trace of their origins. They strike specks of dust coming from all possible directions. The impacts are distributed randomly.

6. According to Heisenberg's uncertainty principle, the possibility of determining the energy of a system is related to the amount of time during which the system is under study. A temporal period of 10^{-43} seconds (Planck's time) corresponds to an indetermination in energy of 10^{28} electron-volts (Planck's energy), (equivalent to the kinetic energy of a tennis

ball hit by a champion). At such energies, the quantum fluctuations of the gravitational field make impossible any predictions concerning the future of the system. In other words, it is physically impossible to tell the hour with any greater precision than Planck's time. This is not a technical limitation linked to the state of our current technology but rather a limit imposed by nature (as we now know it, at any rate).

7. Recent calculations by Jacques Laskar of the Bureau des longitudes de Paris show that a mere uncertainty of 15 meters in measuring the position of the Earth in its orbit (a relative uncertainty of 10^{-11}, or one part per hundred billion) is sufficient to make impossible any prediction of the Earth's orbital position in a hundred million years. His work shows as well that, relatively speaking, the orbits of Mercury, Venus, and Earth are chaotic.
8. It is not always necessary for there to be a large number of bodies to create chaotic conditions. In some instances, very few are needed. The preceding note provides some examples.
9. A number of lively terms and expressions have been introduced in this chapter. Edward Lorenz discovered and named the "butterfly" effect, in a paper titled "Predictability: Does the Flap of a Butterfly's Wings in Brazil Set Off a Tornado in Texas?"

   There is also Feigenbaum and his "baker's transformation"; Mandelbrot and fractal geometry; David Ruelle, E. Takens, Michel Hénon and "strange attractors." Read: *Chaos*, by James Gleick (Viking Penguin, 1987).

   We should also mention the name of Ilya Prigogine, in connection with his "dissipative structures." I discovered with some surprise that there was a lively controversy surrounding his name. Curious to know why, I talked personally with a number of specialists.

It's not his questioning of the absolute determinism of classical physics that is being contested but rather the actual contribution of the Brussels school in this enterprise.

It's generally recognized that Prigogine's work on the thermodynamics of irreversible processes paved the way for this new approach. But it's further claimed that where crucial progress on the theory of dynamic systems is concerned, the names cited earlier must be given particular credit (see "Les ondes chimiques" by Christian Vidal, *La Recherche* [December 1989]).

In their joint volume, *Order out of Chaos* (Bantam, 1984), Prigogine and Isabelle Stengers revealed to a wide public how time had made its return to the world of science. This book has had a marked influence on current thinkers.

10. The naturalist Buffon, who was little respected by his physicist colleagues, seems here to have outdistanced them. In his *Histoire naturelle*, he wrote in 1783: "Everything proceeds because, in time, everything intersects, and in the free expanse of space and in the round of movement, all matter is stirred, all form bestowed, all figures traced; thus everything approaches or grows more distant, everything becomes one or breaks apart, everything attracts or repels, everything is created or destroyed by connecting or opposing forces that alone are constant, and, poised without annihilating one the other, give life to the Universe and make of it a theatre whose scenes are always new and whose objects are constantly being reborn." (*Histoire naturelle des minéraux*, 1783, tome 2, p. 107).

Chapter 6: Thermodynamics and the Flight of Butterflies

1 For more information on the role of photons in entropy in the universe, see *The Hour of Our Delight.*

2. In a physical system where gravitational energy dominates other forms of energy (a star, for example) *the isothermal state is not the state of greatest entropy*. The gaseous mass increases its entropy in collapsing in upon itself. This collapse raises the temperature in the center, creating an ever-widening temperature gap between the center and the surface. Applying the *virial theorem*, we can relate the quantity of gravitational energy freed in the form of photons to the increase in internal energy of the star, that is, the heightening of the internal temperature. It's thanks to this particular property of gravitational force that the universe emerged from its isothermal state, and that cosmic energy could be put to use.

3. What we have here, of course, is an analogy for teaching purposes. There is no suggestion that science, in learning to speak this language, would be in a position to exhaust reality. This subject was raised in Chapter One.

4. Following on the work of Stephen Hawking, we now think that black holes are not really black. Thanks to a quantum effect, they possess a thermal radiation whose temperature is inversely proportional to their mass. This radiation could, in the distant future, become the most important source of information in the universe (see *The Hour of Our Delight*).

5. We can find some superb examples of this in the work of Claude Nuridsany and Marie Perennou, *Voir l'invisible* et *Photographier la nature* (Hachette).

6. The important factor here is *the stability of structures*. Reactions between different particles tend to favor structures that are stable. As the universe cooled, the chaotic monotony of the landscape gave way to a great variety of atomic and molecular forms. Their appearance was made possible by the fact that they were more stable than the constituents of which they were made up.

Physics teaches us that if the search for stability were pushed to the limit, the universe would be plunged again into monotony. At a high temperature, the most stable state is the soup of particles disclosed to us by our observations of the distant past. At our temperatures, the most stable state is iron. If the evolution of cosmic matter had been determined entirely by the quest for stability, our universe would contain nothing today but atoms of iron. The initial monotony of elementary particles would have been succeeded by a monotony of iron.

The appearance of a variety of atoms and molecules is due to the activity of nuclear and electromagnetic forces trying to produce structures that are more and more stable. This effort has been impeded by the onset of a disequilibrium, without which this diversity would have been reduced to a new monotony.

This disequilibrium is extremely fertile. Not satisfied with neutralizing the natural tendency of forces to reimpose a monotonous universe, it opens the door to what is new and original. Under its aegis, the results of interactions are largely unpredictable. Preexisting structures (atoms, molecules) can combine to produce new entities, which in turn can play the same game.

Chapter 7: Butterflies and Time's Arrow

1. A simple example will show that the transition from less probable to more probable does not necessarily make for anything new.

   Here is a filmed sequence for which, in accordance with our game, we must identify the correct transition. We film a game of dice. In the first image, there are two dice on the table. We add up the figures. We throw the dice. We locate them on the table. We do the sum again.

   Imagine that the first sum is 2 and the last one 7. We know that the figure 2 can only result from 1 on the first die and 1 on the second. We can get the figure 7 in six different ways: 1+6, 2+5, 3+4, 4+3, 5+2, 6+1. The sequence from 2 to 7 is more probable than the sequence from 7 to 2. The true sequence is probably that which begins with 2 and ends with 7.

   This sequence illustrates that the transition from the improbable to the probable in no way implies that there will be anything new. The six final states were known in advance. We knew that one of the six would turn up, but we didn't know which.
2. More exactly, the drop of ink has no more chance of reforming on the basis of the initial data resulting from the previous calculation than if we were starting from any other selection of initial data chosen at random.
3. The objection might be made that if our knowledge of the initial conditions is too limited, this is the fault of the computer and not the drop. The ink molecules "know" their positions and their speeds. They would know how to find their way back to the drop, even if the computer doesn't.

   Let us say, first of all, that computers are easier to control than glasses of water. A good computer, once at work, can complete its calculation in peace, without any outside disturbances intervening. Molecules in a glass are not so secure. The smallest jolt — a truck in a nearby street, an airplane in the sky, even a current of air in the atmosphere of Neptune — is enough to create a tiny vibration, all it takes to disturb the trajectories. The "butterflies" always have the last word.

Moreover, let's not forget that molecules are subject to quantum indeterminacy. There's a natural limit to the simultaneous determination of their positions and speeds. This limit does away with the notion of a perfectly defined trajectory and is enough to eliminate all hope of reforming the drop. We are in the world of our third clock, limited by "Planck's time."

4. It is now acknowledged that physical equations, both classical and quantum, are reversible in time. This means that if we replace the variable "t" in their mathematical expression with "-t" (reversing the direction of time) that would not change the terms of the equations. What we've discovered is that the invariance of the *equations* in relation to time does not guarantee the invariance of the *solutions to the equations* in relation to the reversal of time's arrow.

In other words, even if the equations give us the impression that the past and the future are mirror-images of each other, this impression is not supported by the solutions to the equations, especially when we extend them beyond their predictive horizons.

Which brings up the question once more: if for short periods the past resembles the future, how is it that they can be distinguished over long periods?

Prigogine has tried to answer this question by questioning

the temporal invariance of the equations themselves. According to him, traditional equations of quantum physics — invariant in relation to time's reversal — could be replaced by a new set of equations, just as viable, but which are not invariant in relation to a reversal in time. But for the present, this has yet to be proven.

In this chapter I've tried to show that in fact the past and the future are different on all time scales, short or long. I take up this question in more detail in Appendix B.

5. The fact that the billiard table is not situated outside, under the sky, does not affect the argument. The sheltering structure then becomes the intermediary by which the infrared photons are exchanged with the universe.
6. The concept of predictive horizons applies just as much to a world in equilibrium as to one in disequilibrium. But their "liberating" effect is then neutralized.

   Let us fix our attention on the future of a group of particles. Our knowledge of their positions and their speeds at a given moment allows us to foresee their behavior over a certain period whose duration is determined by the predictive horizon. The configuration they will adopt beyond this horizon escapes us.

   But in a state of equilibrium the configurations average out. The overall result is calculable at any instant by a simple application of the formulas for equilibrium.

Chapter 8: The Artisan of the Eighth Day

1. In the autumn of 1989 the COBE satellite, equipped with appropriate detectors, was put into orbit to study, with greater

precision, the properties of this famous fossil radiation. The first results stupefied the astronomical community. The radiation matched, almost exactly, the expectations of the Big Bang theoreticians. It supported the picture of an ancient universe that was perfectly disorganized, perfectly "thermalized." In April 1992 the detection was reported of a faint granularity in this radiation.

2. I am consciously and deliberately using ambiguous language in discussing nature and human beings. Of course, humanity cannot be disassociated from the natural context in which it finds itself. It's as much a part of nature as stones or warblers. But by removing it provisionally from its setting, we can shed light on the situation that interests us. As long as we reintegrate it afterwards. We'll return to this theme in Chapter Nine.
3. It remains an open question whether these new instruments will transmit feelings as intense as those conveyed by the classical orchestra. Only the future and the creators themselves will provide an answer.
4. For the Greeks, the word "cosmos" meant the universe, order, and beauty in the universe. Our word "cosmetic" derives from this.

Chapter 9: A Smokestack in the Countryside

1. In *Timon of Athens* (Act IV, scene II), Shakespeare presents us with his personal vision of the animal world. Peaceful coexistence and harmony in an earthly paradise do not seem to be the order of the day.

   "If thou wert the lion, the fox would beguile thee: if thou

wert the lamb, the fox would eat thee: if thou wert the fox, the lion would suspect thee, when peradventure thou wert accused by the ass: if thou wert the ass, thy dulness would torment thee, and still thou livest but as a breakfast to the wolf: if thou wert the wolf, thy greediness would afflict thee, and oft thou shouldst hazard thy life for thy dinner: wert thou the unicorn, pride and wrath would confound thee, and make thine own self the conquest of thy fury: wert thou a bear, thou wouldst be killed by the horse: wert thou a horse, thou wouldst be seized by the leopard: wert thou a leopard, thou wert german to the lion, and the spots of thy kindred were jurors on thy life: all thy safety were remotion, and thy defence absence. What beast couldst thou be, that were not subject to a beast?"

2. This being said, the reality is not so simple. On the scale of biological evolution for species, our observations apply to very short time periods.
3. The social behavior of termites illustrates nature's apparent lack of interest in individuals. The "law of the termitary," which is focused entirely on the good of the society, ignores the person-termite. It is allowed no initiative, and it will be mercilessly sacrificed if the group's salvation requires it.

   How does modern society fare in this respect? The year 1984 revived memories of George Orwell's novel. Over the years, this date had been invested with dire connotations, a vague whiff of the "end of the world." It was associated with the idea of a "petrified" human species. A kind of death for humanity. Not in the sense of *rigor mortis*, but affecting all that makes life worth living: freedom, creativity, fantasy.

   As I remember it, Orwell's novel, along with Aldous

Huxley's *Brave New World*, gave us a foretaste of our greatest threat: the termitary. The specter of a strong, impersonal state, armed with techniques for controlling individuals. A secret police, brainwashing psychologists, and genetic manipulators working together to persuade each citizen that "*Big Brother is watching you*" and that all hope for freedom is not only futile but harmful.

What of these predictions at the beginning of the 1990s? On the international level, happily, Orwell's dark presentiments do not seem to have been justified. Many European and South American countries have rejected their dictators. Huxley was more clairvoyant. The specter of the termitary is haunting the democracies themselves. The growing proliferation of numbers, cards, and papers required in order to have the *right to exist* gives the state a measure of control over each individual. Enticed by the slogan "The state wants your good," each citizen is invited to take shelter under its protective wings, to become "socially secure," childlike, and without responsibilities. Crucial decisions for our future are often taken by impersonal and incompetent bureaucracies. Here again, the future of creativity seems very much compromised.

If I'm optimistic, it's because of the dissidents. To paraphrase the Bible, we may say that "the fear of the state is the beginning of wisdom." It carries within it a distrust of centralization. It favors individual initiative and small-scale ventures. "*Small is beautiful.*"

The long-term solution, our hope for life and survival, can only derive from personal development. The strong states find their recruits among those individuals without character, susceptible to propaganda and to the power of mythic images. It's the myth of the "leader" (the "führer") that started millions of Hitler youth along the road to butchery.

But it will be a long struggle. The arms are not external but internal. At every level, educators must become involved, not to furnish new pawns for the state, but to promote the discovery of the self and its potential. Free and responsible citizens are our best defense against the termitary.

4. Throughout these pages, we've talked about the stormy relationship between *man and nature*. It would be more appropriate, perhaps, to speak of the relationship between *intelligence and nature*. On our planet, it's the human being whose intelligence has outstripped that of any other living species. Elsewhere in the universe, the situation might be different.

   Where survival strategies are concerned, intelligence wins first prize. It's conceivable, therefore, that it may have developed among certain living species that make up the fauna of hypothetically inhabited planets. Who can say whether, in the planetary entourage of the star Bêta Pictoris, hominoids may not be the guinea pigs of guinea pigs? Perhaps, in the neighborhood of Capella, rats show them off in their traveling circuses. If so, we may well imagine that these species, having attained a level of intelligence equal or superior to our own, are being faced with the same sorts of problems as ourselves.

5. *Le Sexe et l'Innovation*, by André Langaney (Le Seuil, 1987), coll. "Points Science."
6. "What business is this of his? Since when do astrophysicists feel they're qualified to give us their opinions on abortion?" I think, personally, that the problem is sufficiently important and difficult enough that everyone should express an interest and give his or her point of view.

Chapter 10: A Steeple over Purple Beeches

1. Charles Baudelaire, *Les Fleurs du Mal*, translated by Richard Howard (David R. Godine, 1982).
2. "For the old Gods came to an end long ago. And verily it was a good and joyful end of Gods!

   "They did not die lingering in the twilight — although that lie is told! On the contrary, they once upon a time — laughed themselves unto death!

   "That came to pass when, by a God himself, the most ungodly word was uttered, the word: 'there is but one God! Thou shalt have no other Gods before me.'

   "An old grim beard of a God, a jealous one, forgot himself thus.

   "And then all Gods laughed and shook on their chairs and cried: 'Is godliness not just that there are Gods, but no God?'" Nietzsche: *Thus Spake Zarathustra*.
3. Friedrich Nietzsche, from *Beyond Good and Evil*, translated by Walter Kaufmann (Vintage Books, 1989).
4. For an intelligent, balanced critique of sociobiology, I recommend Stephen Jay Gould's book *Ever Since Darwin*, (Norton, 1979). Gould acknowledges the useful contributions of this discipline while discarding its exaggerated pretensions.
5. E. O. Wilson, in *The Nature of Man*.
6. Many believers, from different religions, have tried to show that their traditional teachings anticipated scientific research. Such efforts have as their corollary that their religion is the only valid one, and before its appearance on the scene humanity was in a state of suspension, waiting for the truth that would free it . . .

Recently, in the Islamic world, scholars have gone to great pains to demonstrate that the Koran contains many scientific truths later confirmed by modern research. Thus Mohammed was doubly prophetic: a prophet of God, but also a prophet of science.

I recently had the opportunity to discuss this question with a Muslim intellectual in Algeria. This meeting was very enlightening for me.

Above all, I wanted to understand why such an enterprise was so important to this learned and cultivated man. "The teachings of the Koran have their own value, even if Mohammed is not the scientific precursor you want him to be," I said, "so why are you so attached to this cause that's so difficult to defend?"

"The Koran makes religious, political, and scientific pronouncements all together. If its science is true, the credit it receives reflects on the other pronouncements as well."

Implicit in this reply, it seemed to me, were his own doubts concerning the message of his religion. He needed a double protection. To be really believable, Mohammed's message had to incorporate both wisdom and scientific truth.

"And if you discover that the Koran contains scientific errors, what will you do?" "Impossible," he said. "The Koran cannot be wrong, because it is the word of God." I wondered if he was truly convinced.

7. In writing these lines, I am not forgetting the negative aspects of religion, examples of which may be found throughout human history. The creation of religious institutions generally involves a distortion of the message. It introduces questionable elements, such as the taste for power.

8. Bertrand Russell, *Religion and Science* (Henry Holt and Company, 1935).
9. In *The Critique of Pure Reason.*
10. Friedrich Nietzsche, from *Beyond Good and Evil, op. cit.*

Further Reading

Careri, Giorgio. *Order and Disorder in Matter.* Translated by Kristin Jarrat. Menlo Prk, California: Benjamin/Cummings, 1984.

Crescenzo, Luciano de. *Pythagore Superstar.* Paris: J.C. Lattès, 1985.

Davies, Paul. *The Cosmic Blueprint: New Discoveries in Nature's Creative Ability to Order the Universe.* New York: Simon and Schuster, 1989.

Dolto, Françoise. *Autoportrait d'une psychanalyste.* Paris: Le Seuil, 1989.

Gleick, James. *Chaos: Making a New Science.* New York: Viking Penguin, 1988.

Gould, Stephen Jay. *Ever Since Darwin: Reflections in Natural History.* New York: Norton, 1979.

Humbert, Elie. *Jung.* Paris: Editions universitaires, 1983.

Kline, Morris. *Mathematics: The Loss of Certainty.* London: Oxford University Press, 1980.

Margulis, Lynn, and Dorion Sagan. *Microcosmos: Four Billion Years of Evolution from Our Microbial Ancestors.* New York: Simon and Schuster, 1991.

Ornstein, Robert, and Richard Thompson. *The Amazing Brain.* Boston: Houghton Mifflin, 1986.

Penrose, Roger. *The Emperor's New Mind: Concerning Computers, Minds and the Laws of Physics.* London: Oxford University Press, 1989.

Prigogine, Ilya, and Isabelle Stengers. *Order Out of Chaos: Man's New Dialogue with Nature.* New York: Bantam, 1984.

Robin, Jacques. *Changer d'ère.* Paris: Le Seuil, 1989.

Winnicott, D. W. *Playing and Reality.* London: Tavistock, 1982.